W9-BNP-530

Lily
one in a million!

... A MIRACLE
OF SURVIVAL

Laura Hamilton

Hubble & Hattie

The Hubble & Hattie imprint was launched in 2009, and is named in memory of two very special Westie sisters owned by Veloce's proprietors. Since the first book, many more have been added, all with the same underlying objective: to be of real benefit to the species they cover; at the same time promoting compassion, understanding and respect between all animals (including human ones!) All Hubble & Hattie publications offer ethical, high quality content and presentation, plus great value for money.

More great books from Hubble & Hattie –

www.hubbleandhattie.com

First published March 2018 by Veloce Publishing Limited, Veloce House, Parkway Farm Business Park, Middle Farm Way, Poundbury, Dorchester, Dorset, DT1 3AR, England. Tel 01305 260068/Fax 01305 250479/email info@hubbleandhattie.com/web www.hubbleandhattie.com ISBN: 978-1-787111-47-9 UPC: 6-36847-01147-5 ©Laura Hamilton & Veloce Publishing Ltd 2018.
All rights reserved. With the exception of quoting brief passages for the purpose of review, no part of this publication may be recorded, reproduced or transmitted by any means, including photocopying, without the written permission of Veloce Publishing Ltd. Throughout this book logos, model names and designations, etc, have been used for the purposes of identification, illustration and decoration. Such names are the property of the trademark holder as this is not an official publication.
Readers with ideas for books about animals, or animal-related topics, are invited to write to the publliser of Veloce Publishing at the above address. British Library Cataloguing in Publication Data – A catalogue record for this book is available from the British Library. Typesetting, design and page make-up all by Veloce Publishing Ltd on Apple Mac. Printed in India by Replika Press.

Contents

Dedication &
Acknowledgements

Dedication

To the memory of Jim, my wonderful husband; my soul mate.

And to the memory of Margaret, my fabulous mother.

Acknowledgements

A very big thank you to –

Jim, my wonderful husband and soul mate, for enriching my life with our amazing marriage. You made writing this book and especially its preface intensely personal and meaningful.

My mother, for always believing in me, for making me learn to touch-type the summer I turned fifteen, for fostering my love of writing by your example, and for knowing all along that Jim and I were meant for each other.

Elizabeth (Lizzy): For all you did for Lily in her first weeks; sitting in the whelping box and aspirating milk from her nose. You gave me time to look after Mini, Liana, and the thriving puppies. I'm grateful you still answer when I call you Lily by mistake! (Which is fortunate because I'm still doing that!)

Marcus, Elizabeth, and Shaun: For your technological wizardry in helping me get Lily's book to the publisher. It may not seem like magic to you, but it does to me.

The ladies in the Lockswood Women's Institute: You found my continuing challenge of caring for Lily so heart-warming that you felt I should write a book about her. I promised I would if she made it through her first year. After her first birthday, you asked 'Have you started Lily's book?' I admitted I hadn't. But you kept asking, every month, every Lunch Club, every meeting: had I started Lily's book, yet? Finally, I felt I couldn't keep disappointing you, so, three days before Lily's second birthday, I began writing her story, and you were delighted. Thank you for that year of persistent persuasion. It was definitely worthwhile: Lily's story needed to be told.

Dr Pete Wedderburn and Dr Lucy Atkinson: The research you did on your own, just to learn about Lily's condition and how rare it is, certainly validates the title of this book. Thank you so very much.

Dedication & Acknowledgements

Jackie Foster: For writing the foreword, which reflects how caring a breeder you are. I really appreciated your support when I chose not to euthanise Lily.

Kevin Atkins at Hubble & Hattie: For working his magic on my photographs. Thank you so much for resizing those photos which were really too small but too important not to include in Lily's book. Each picture – or, in this case, each photo – is worth a thousand words; you've helped immeasurably to enhance the telling of Lily's story.

And, of course, Jude Brooks, my wonderful publisher. Jude, you are absolutely amazing. The title you thought of for Lily's book couldn't be better. No exaggeration, for Lily *is* one in a million. No fabrication, for she *is* a miracle of survival. I never would have thought of writing the preface and the diary of Lily's day without your suggestion.

Your other suggestions about having a foreword by a distinguished Golden Retriever breeder, and testimonials by veterinarians about the rarity of Lily's condition were inspired. You were very understanding about my technological shortcomings, too, like when I transferred all the photos and captions to the wrong email address (wherever they went, they were never downloaded), and sent you a file so compressed that it wouldn't uncompress! Because of you, I'm having my very first book published.

And you've given Lily the opportunity to share her story so that other animals with life-threatening conditions may be given the chance to survive and live life to the full. So, Jude, thank you very, very much.

Laura Hamilton
Southampton
Hampshire, England

Foreword

When Laura first contacted me, I felt immediately that she would be a good 'mother' to one of Figgy's puppies. When we met I was absolutely certain. For me, as a breeder, it is essential that I meet each prospective owner to understand their home environment, their other commitments, and their appreciation of the time they will have to invest to nurture and train a puppy to ensure they grow up to become a sociable and well behaved dog. Laura passed with flying colours!

Jackie Foster with (from top) Pilot, Lily, and Honeybury Golden Retrievers Buffy, Bumble, and Rhubie.

I discovered that, like me, she had been a schoolteacher. She had raised and trained a prize-winning German Shepherd when she lived in the USA, and now she was determined to devote her patience, energy, time and love to bringing up one of my Golden Retriever puppies.

As Laura lived nearby, I was able to watch Pilot grow up – an added bonus. We enjoyed many walks by the river where Pilot was able to play with my girls under the watchful eye of her mum, Figgy!

I have owned Golden Retrievers since 1981, and I am a certified member of the Kennel Club's Assured Breeders Scheme: the only organisation given UKAS accreditation to certify dog breeders. I enjoy keeping in touch with all my puppies at their new homes, and encourage owners to contact me whenever they have any questions or worries about their dogs, and to advise me of their progress.

I have always looked forward to receiving updates from Laura about Pilot, Lily, and her big brother, Bentley, and I hope I have been able to provide her with advice and reassurance over the years.

With this book we are given an insight into the devotion and dedication required to find the ways necessary to provide Lily with a full life, which is rightly described as a miracle.

Lily is a wonderful dog who is giving strength, confidence, joy, and love to those she meets, both young and old. As Pets As Therapy dogs, both she and Pilot are continuing to make new friends, and give pleasure and assurance to so many people.

She is good-natured, kind, calm, and confident. She is a Golden Retriever. This is Lily's book. And it is Laura's, too.

Jackie Foster
Honeybury Golden Retrievers

Preface

Lily wasn't even fifteen minutes old before I knew something wasn't right about her. I couldn't imagine what was wrong with her, and little did I know that she'd been born with a condition so unusual it makes her one in a million dogs.

When people ask about Lily, I describe her disability, how rare it is, the care she needs, the challenge of looking after her, and her work as a Pets As Therapy dog. "It's a miracle she's still with you," "You've worked a miracle with her," and "What Lily does for others is a miracle," are what they usually say, but, until I heard such comments, I was so focused on helping Lily to survive and have a really good life that it never occurred to me there was anything miraculous about her. To me, a miracle was always something wonderful caused by God, but, because of Lily, I've learned that a miracle is also something surprising, unexpected – and seemingly impossible.

With my new understanding of miracles, I've realized that Lily really *is* a miracle dog, and, not only that, but she's my second 'miracle' dog. Ty was my first.

I homed Ty years ago when I was living in the State of Washington. Having wanted a German Shepherd puppy for a while, I responded to a 'Pets for Sale' ad for German Shepherd puppies in a local newspaper. I never questioned why the breeder still had three, 4-month old puppies for sale, although I should have: it was because those who knew anything about German Shepherds had the good sense not to buy any of these particular pups ...

When I took Ty to the vet for a check-up the next day, the vet told me bluntly that he was a poor quality dog, who had never been socialised. Insecure, Ty was going to bite from fear if I didn't give him intensive training – and sooner, rather than later. I began training Ty the very next week. Luckily, he loved learning, and was very bright: doing really well in the American Kennel Club's obedience trials for –

• Companion Dog (on-leash and off-leash heel, and heel a figure 8; stand for examination; recall; one minute long sit; three minutes long down)

• Companion Dog Excellent (figure 8 off-leash heel; drop on recall; broad jump; retrieve on flat; retrieve over a high jump; three minutes long sit and five minutes long down, with handler out of sight; voice and hand commands)

• Utility (hand signals only for stand, stay, down, sit, come; scent discrimination using 12 dumbbells, six metal and six leather; directed jumping; directed retrieving; moving stand and stay for examination by judge; return to handler on command)

Ty became so confident as he amassed obedience trophies that he was even considered to have the right temperament to undergo protection training: something we enjoyed enormously.

To my astonishment, the American magazine *Dog World* gave Ty its 'Award of Canine Distinction' certificate, and recorded his name in its *Album of Great Dogs of the Past and Present*.

Ty's transformation from a fearful, unsocialised puppy to an award-winning dog was a miracle. Surprising. Unexpected. And seemingly impossible.

Ty and Lily have both brought miracles into my life. And Jim brought another.

I met Jim in Toronto, Canada, when I was eight years old. I decided to marry him. He was twenty-five, British, a pilot with Trans-Canada Air Lines (TCA), and boarding at my grandmother's house with two other TCA pilots: my Uncle Trevor, and Ed, also British.

It took me until I was ten to persuade Jim to have a picnic with me in the park. We ate peanut butter sandwiches.

When I was twelve, Jim broke my heart when he and Ed left Canada to return to Britain to fly for British European Airways. I was devastated. I had to grow up without him.

Twelve years passed before I saw Jim again, by which time, he'd faded from my memory. During my summer vacation from teaching English, my mother and I, at her request, were to have his family home as a base while we visited London, Wales, and Paris.

Jim was flying when his wife met us at the train station, and drove us home. I was upstairs unpacking when I heard his car in the drive. Curious to see the man my eight-year-old self had intended to marry, I waited at the top of the staircase as he entered his house. The moment I saw him, I felt a supernatural certainty that I would marry him!

My mother and I spent little time in Jim's house as we travelled around, and, even when we were there, Jim was usually flying. Back home in Canada, I had no reason to think I'd see him again, and felt absolutely distraught at the prospect.

Two years passed, and I was living in my own apartment. One Friday in early October, I went to bed, thinking of what I had to do over the weekend: mark papers, and write and illustrate a children's story for a newspaper. Though Jim was often in my thoughts, he definitely wasn't that night.

At 6:10 the next morning, I was jolted awake by a strange sensation, with one thought in my head: 'It's not Jim.' The experience confused and shook me, and didn't make sense. Why in heaven's name had Jim come into my mind out of the blue? And what did it mean?

I was still feeling anxious and upset when I turned on the radio for the seven o'clock news, to hear that a British European Airways plane had crashed in Belgium at the very moment I had jolted awake. Tragically, everyone on board had been killed. But then the horror became personal: the captain had been Ed, who had lived with Jim and my uncle at my grandmother's. We had all liked Ed, and my last memory of him was when he treated my mother and me to afternoon tea in Windsor.

The investigation into the crash exonerated Ed of any blame: he could not have saved the Vanguard. The force of pressurization at 19,000ft (5790m) had caused the corroded rear bulkhead to fail. The tail broke away.

Jim had flown that same Vanguard only two weeks earlier.

I was overwhelmed at how the cryptic thought 'It's not Jim' now made sense. It wasn't Jim who had been the captain on that plane. Had I somehow been told this because my destiny was to be with him? If so, it would take a miracle, for we were not in touch.

Eight years later, convinced Jim and I would never be together, I was living near Seattle in the wrong marriage, when my mother phoned to say Uncle Trevor had told her that Jim – now a Senior Base Training Captain with British Airways – sometimes passed through Seattle, en route to Vancouver to train BA pilots using another airline's simulator there. She suggested I contact him; renew the family acquaintance. So I did. And he responded.

Eleven years after seeing Jim in England, I drove alone to Sea-Tac, Seattle's airport, where we spent two hours together with another BA training captain and his wife. Jim was now divorced, but I was married. Too soon, he and the other captain left for Vancouver, the captain's wife flew to London, and I returned to the wrong marriage.

Jim and I at our 'peanut butter picnic park' in Scarborough, Ontario, photographed by our son, Marcus, aged six.

Jim didn't contact me after Sea-Tac, so he wasn't the reason why I filed for divorce, two-and-a-half years later. Washington State law meant my divorce would be final in 90 days.

My mother urged me to write to Jim and tell him, so I did, and he phoned me on Christmas Day while I was bathing Ty.

After that, we phoned or wrote to each other every day, and Jim asked me to come to England in the Easter holidays to marry him!

I secured early release from my teaching contract, found loving homes for my dogs and cats, left my house, car, friends, and belongings in Washington, and, without a moment's doubt, boarded the British Airways flight to London.

Jim was waiting at Heathrow for me. For us.

In thirteen years, we had seen each other for just two hours, with chaperones in a busy airport, yet now we were joyously planning our wedding. We just knew we were meant to be together.

We married nineteen days later in a registry office. Over the years, we visited the registrar with our children, and he delighted in saying we were one of his success stories.

Less successful was peanut butter, however. Years after giving Jim a peanut butter sandwich at our picnic, I made peanut butter cookies. Our children enjoyed them, but Jim declined one, smiling: "I didn't have the heart to tell you when you were ten, but I don't like peanut butter."

We had a most amazing marriage for almost 29 years before Jim died from cancer. A dear friend in Ontario wrote on her condolence card 'You are so lucky to have met and married your soul mate. Yours truly was a marriage made in heaven.'

Our marriage was everything a miracle could be. Surprising: Uncle Trevor was

9

certainly surprised when his best friend suddenly became his nephew. Unexpected: by everyone except my delighted mother, three school colleagues and my two closest girlfriends. Almost impossible: definitely. A wonderful and surprising event caused by God. Difficult to deny.

Months after losing Jim, I bought my Golden Retriever, naming her Pilot for him. And then when Pilot had puppies, Lily came into my life, bringing the miracle of her own story.

My wish is that you enjoy Lily's book, and that it brings hope, inspiration, and maybe even a miracle to all of you who have animals who need your help to enjoy a life lived to the full.

Visit Hubble and Hattie on the web: www.hubbleandhattie.com
hubbleandhattie.blogspot.co.uk
• Details of all books • Special offers • Newsletter • New book news

10

Puppy love

Having a litter of Golden Retriever puppies was going to be wonderful; everything my daughter, Lizzy, and I dreamed it would be. We imagined ourselves watching newly-born puppies softly snuggled up to their mother, my dog, Pilot, making soft murmurings of contented nursing before falling asleep.

I knew how lovely puppies can be, for, years earlier, when I was living in Washington State in the United States, I bred my German Shepherd, Maggie, twice: the first time when she was two years old, and the second time when she was four years old. Maggie produced two litters of purebred pups.

I'd had no experience of what the actual whelping was like with either of Maggie's litters, because both times she'd had her puppies by Caesarean section, although neither time had been an emergency. Maggie had had primary uterine inertia: in layman terms, it meant she had just never gone into labour.

The first time that this happened was possibly because she was carrying just two puppies. It was thought that the hormonal changes that help start labour are triggered by rising levels of the stress hormone cortisol, which the developing puppies produce, and maybe Maggie's two puppies weren't enough to produce the cortisol level needed to get her going. But, even if they had been, one of the puppies was twice the size of a normal German Shepherd puppy so she would have needed a Caesarean anyway, the vet told me. He was simply too big for Maggie to have pushed him into the world.

The second time Maggie failed to go into labour was probably not because of insufficient cortisol levels – she was carrying eight puppies: plenty, I would have thought, to raise cortisol levels sufficiently. However, maybe Maggie's size was a reason why she needed a C-section, as she was small for a German Shepherd. Anyway, she never had another litter as I had her spayed after her next season.

Both puppies in Maggie's first litter survived, but one of the eight from the second litter died sometime during the first night. I was devastated. I'd never expected a puppy to die, and was all for having the vet do an autopsy to find out why. He dissuaded me, explaining that it isn't unusual to lose 25-30 per cent of the puppies in a litter. Nevertheless, losing the puppy was a dreadful shock to me, and I was very sad.

A few years after that experience, which taught me so much, I moved to England to marry Jim.

Jim loved dogs as much as I did. When he was eleven years old, whilst visiting his grandparents and cousin in Rock Ferry by himself, he adopted a stray dog, naming him Pete. Jim took Pete home on the train 200 miles to Glasgow where his father was stationed. His mother, somewhat surprised, welcomed Pete with a steak. Jim used to entertain me with stories about Pete, and how he was a great companion for him, an only child, and his mother while his father, a captain in the merchant marine, was away at sea for months at a time.

Lily – one in a million!

Jim and I would have dearly liked a dog, but we knew having one wouldn't have been right, as we weren't home enough. Jim was a captain with British Airways, and I accompanied him on quite a few trips. Even after our children were born, we flew together as a family fairly frequently. During holidays and half-term breaks, we were gone for weeks at a time. We couldn't have given a dog the life every dog deserves, lived to the full with love, attention, companionship, and training.

After almost 29 years of an amazing marriage, Jim passed away from cancer. Marcus, our son, and Lizzy, our daughter, now independent adults, were living away from home, so I was alone.

Months after Jim's death, Lizzy decided I needed a dog for company. And she knew exactly what breed of dog I should have: a Golden Retriever. She had always wanted one, ever since seeing the photos and reading about Golden Retrievers in a copy of *The Ultimate Dog Book* that I had won at a church raffle when she was about eight.

Apparently, she tells me she used to leave the book open at the two-page Golden Retriever spread at every opportunity, hoping her dad and I would notice. To her disappointment, we never took her hint on board.

If Lizzy hadn't felt so passionately about Golden Retrievers, a very worthwhile alternative to buying a dog would be taking in an animal from a rescue centre, and giving him or her a much-needed forever home. For Lizzy, however, only a Golden Retriever would do, and, searching the UK Kennel Club's website, she discovered a new litter of Golden Retriever puppies. She phoned me at 6am one morning to insist I contact Honeybury Golden Retrievers sooner rather than later, because there were only three puppies in this particular litter.

In the preceding weeks, I had contacted three breeders who had listed new litters on the Kennel Club website, but all had been quickly reserved. I hoped that this time would be different. Unfortunately, when I phoned Jackie Foster, of Honeybury Golden Retrievers, she told me that two of the three puppies were already reserved, and she was keeping the third puppy herself!

Jackie could tell I was very disappointed, so told me that she did have another litter of ten puppies, who were only five days old; too young to put them on the Kennel Club website.

I'd finally found a breeder who still had puppies, and three of them were cream-coloured girls: exactly what I was looking for.

Jackie began to ask me a lot of very searching questions. Why did I want a Golden Retriever? What experience did I have of owning a dog, especially a big dog? Would the dog be left alone a lot? Were there very young children in the house? Would I be able to give a dog the necessary exercise and training?

It became obvious to me that Jackie's priority was not to sell me one of her puppies, but to ensure that the puppy would be going to the right home. She certainly had the pups' lifetime welfare and needs uppermost in her mind, so it was a huge relief when it became apparent that Jackie considered I might be the right kind of person to become a 'mother' to one of her puppies.

She still had to meet me to make sure, however, so, would I like to come and see the puppies? she asked. Would I? She didn't have to ask me twice, and, incredibly, she lived just six miles from me.

When I visited Honeybury Golden Retrievers, I was completely won over.

Among a litter of ten pups were the three cream-coloured girls, and, after scrubbing my hands with the anti-microbial skin cleanser that Jackie provided, I held each of them, knowing that one would eventually come home with me. Eight weeks later, I brought home the puppy I had chosen, the gorgeous little girl whom I named Pilot in Jim's memory.

Within a year, Pilot and I had moved 80 miles away to the south coast, and lived a three-minute walk away from Lizzy. After moving, I missed Jackie, who had become such a good friend, but we have certainly stayed in touch. And now that Pilot was almost two years old, Lizzy and I were sharing the dream of breeding from her, but, before that could happen, Pilot had to be screened to ensure she did not have any condition that would cause suffering to her puppies. In this respect, Jackie was indispensable in alerting me to absolutely everything that had to be done beforehand.

Pilot had to be at least year old before undergoing three tests prior to mating –

• her hips had to be x-rayed and scored by the British Veterinary Association (BVA) for evidence of hip dysplasia[1]
• her elbows also had to be x-rayed and scored by the BVA for evidence of elbow dysplasia[2]
• her eyes had to be examined by a veterinary ophthalmologist. Golden Retrievers are prone to the hereditary conditions Generalized Progressive Retinal Atrophy, Central Progressive Retinal Atrophy, Cataracts, and Multifocal Retinal Dysplasia[3]

Pilot passed all of the tests with flying colours.

Pilot's carefully chosen Golden Retriever mate – cream-coloured, like her – also had all of the certificates required, so was completely clear and suitable for breeding. Having two clear parents is no guarantee that their puppies will also be clear, but it does greatly increase the likelihood of this.

Eleven days after Pilot came into her second season, just before she was two years old, she had the first progesterone blood test to determine her hormone levels, which would indicate whether or not it was time to breed her. It wasn't. She had a second progesterone blood test five days later, which showed her to be so ready

[1] A dog with hip dysplasia – abnormal development of the hip socket – may suffer crippling lameness and painful arthritis. Hip dysplasia scores differ breed-to-breed, and range from zero to 106 (the lower the score, the better). The Breed Mean Standard (BMS) for Golden Retrievers is 18. Only those with the current median score of 12 or below should be bred. Pilot's right hip scored 5; her left hip 4, totaling 9: a credit to her conscientious and caring breeder.

[2] Lameness, walking and running with an abnormal gait, holding the elbows at an unusual angle, and not wanting to move around much or play because of joint stiffness and pain are all symptoms of elbow dysplasia. Each elbow is scored from zero (completely clear) to three (badly affected). Pilot scored zero for each elbow.

[3] A Golden Retriever affected by any of these conditions should not be bred. Pilot was clinically unaffected by these inherited eye diseases. However, eye condition can change, and a clear eye certificate is good for just one year. If Pilot had not had puppies within a year of her certificate date, she would have required another examination and a zero score before being bred.

Lily – one in a million!

to be bred that we had almost missed her peak fertile period! Behaviourally and physically, Pilot would be receptive to a male for only a day or two longer.

The morning after I received the result of that second blood test, I took Pilot to the kennel where the sire lived so that she could be mated. Although the mating had been successful, there was no guarantee that puppies would be the result, and I wouldn't know whether or not Pilot was in whelp for a month. Nevertheless, in case she was, it was imperative for her to be vaccinated against the canine herpes virus (CHV), which affects the respiratory and reproductive organs of adult dogs. Adults don't usually show any symptoms, so it's difficult to know whether or not the mother-to-be has the virus. But if she has, and there's a possibility that she is expecting puppies, she should be given the canine herpes virus vaccine, otherwise, the virus could be transmitted to the puppies as they travel down the birth canal, or via the mother's oral and nasal secretions as she cleans and cares for them after birth.

If the puppies should contract the virus, the effects can be awful, with as many as 80 per cent of them bleeding to death during the first couple of days from haemorrhages in their kidneys, liver, lungs, and gastrointestinal tract. Entire litters have been lost because the mother wasn't vaccinated against CHV, which is desperately sad, especially as such a tragedy could have easily been avoided.

The CHV vaccine is given to the mother twice, firstly, during her heat or in early pregnancy, then a booster between one and two weeks before the puppies are due.

After Pilot had her first CHV injection, it was time to apply for a kennel name so that I could register her anticipated puppies with The Kennel Club. As required, I submitted five names, and The Kennel Club gave me my second choice, which I was certainly content with.

Pilot wouldn't show any obvious symptoms that she was in whelp for a month after the mating, when she was scheduled to be scanned. During that long month of waiting, I did a trial run to the emergency vet clinic so I would know where to take her in case she went into labour at night, say, and ran into difficulties.

It never crossed my mind that I might need to make life-and-death decisions about any of the puppies. Why would it? She and her mate had passed all the tests. She'd had the CHV vaccine. She'd continue receiving the best care. Her puppies would develop, having every health advantage. Really, what could go wrong ...?

Four weeks after she was mated, Pilot was scanned, and the scan showed eight puppies: plenty to produce enough cortisol to help start labour.

It was definitely time to check I had all I needed to support Pilot in her whelping, and give her babies the best possible start in life.

At the top of the whelping kit list was a disposable cardboard whelping box, 4ft x 4ft (120cm x 120cm),[4] and at the bottom were baby dog milk substitute powder, bottle, teats, bottle brush and sterilizing tablets to ensure that everything was kept clean.[5]

Fifty days after Pilot was mated, she had the booster CHV vaccination.

A week after that, Pilot was happily wagging her tail while she worked as a Pets As Therapy (PAT) dog at a school in the Read2Dogs Programme; so full of puppies that her PAT jacket wouldn't do up. I joked that Pets As Therapy didn't do maternity jackets. The children looked at Pilot's tummy, fascinated to see puppies moving around.

Sixty days after she was mated, on Bank Holiday Monday, Pilot woke me at

6:15am by thrashing around on my bed. And no wonder: her first puppy was almost halfway delivered!

I ushered Pilot into the whelping box just in time to help deliver Liana, who came out hind legs first (not a particular cause for concern, as around half of all puppies make their entrance this way). Straight after Liana was born, I rang Lizzy: "Get here, NOW!"

On alert for days, Lizzy was with me about a minute after I had put Liana to Pilot to nurse. Lizzy watched, and then whispered what I was already thinking but not wanting to admit: "She can't latch on."

Everything I'd read about what would happen when the puppies arrived told me that they would be eager to nurse. Liana was eager, but something wasn't right. I calmly reassured Lizzy, but was more worried than I let on.

Liana kept falling off Pilot's teat. Every time she did, we put her back on. It was a struggle helping her to nurse, and we ended up holding her mouth to Pilot's teat, hoping she would instinctively know what to do.

In fact, she knew what to do, but she just couldn't do it.

We were still persisting with Liana an hour and five minutes later when puppy number two appeared. This puppy was Bentley, who also came into the world hind legs first, as Liana had.

Golden Retriever puppies should weigh between 350g (12.3oz) and 450g (15.8oz) at birth. Liana weighed 400g (14oz) and Bentley weighed 564g (20oz). Pilot had done well to push him out.

We watched, amazed and delighted, as Bentley latched onto Pilot right away, and began nursing as if in a hurry to grow up into the 38kg (84lb) boy he is today.

Thirty-five minutes later, the third puppy appeared, front legs first this time, we were glad to see. We had another boy: this one weighing in at 400g (14oz).

Twenty minutes after that, at 8:15am, puppy number four was born, front legs first like number three. This one was Lily, initially named Porsche, making her entrance two hours after Liana.

[4]I bought a second 4ft x 4ft (120cm x 120cm) disposable cardboard whelping box when the puppies were four weeks old because they needed more room to move around. I attached this second box to the first, arranging it so that a cardboard panel could block where the boxes connected, allowing pups to eat undisturbed by their littermates who had already been fed or were waiting for food.

[5]I also bought an electric timed heater, anti-microbial skin cleanser, four fleece liners, puppy pads, a dozen terrycloth squares for polishing the newborns (I'm still using these squares to dry off Pilot and Lily after rainy walks), a vinyl pet heat pad with washable cover, and a heat lamp with two spare bulbs. The actual whelping kit contained a medical-grade stethoscope, a 2oz (56ml) bulb syringe/nasal aspirator (a most important item which we used for Liana and Lily more than we ever could have imagined we would!), scales to weigh up to 5000g (11lb), LED torch with batteries, sterile nursing scissors, digital anal thermometer, disposable kidney dish, 2ml (0.07oz) and 5ml (0.2oz) syringes, hand sanitizer, sterile forceps (if a puppy needed forceps for a safe delivery, it would be a vet, not me, using them!), sterile cotton wool, lubricating gel, alcohol swabs, disposable latex gloves, disposable apron, paper towel roll, and a whelping chart for recording each puppy's birth weight, time, presentation, and placenta delivery. Anyone who is put off by having to amass all of these items beforehand is advised to leave breeding to the dedicated, responsible and experienced members of The Kennel Club's Assured Breeders Scheme.

Lily – one in a million!

Lily weighed 406g (14.3oz) – a perfect weight. She began nursing enthusiastically from Pilot: a relief for Lizzy and me, still troubled about Liana who was struggling to stay latched on to her mother. We were so glad Lily had latched on that we never gave a thought to the soft guzzling and gurgling noises she was making. We did notice, however, that although her mouth was firmly on the milk source, her nose, little muzzle, chest, front legs, veterinarian fleece pad – and Pilot – were becoming soaked with milk.

There was little time to wonder what was going on with Lily, though, because the fifth puppy was born ten minutes later, which we thought was quick ... until the eighth puppy put in an appearance less than ten seconds after the seventh!

We breathed a sigh of relief. Pilot's scan had revealed eight puppies, and eight puppies were what we had. We could relax: Pilot's labour was over.

Except that it wasn't ...

An hour and ten minutes later, puppy number nine was born, and twenty minutes after that, puppy number ten entered the world. They must have been hiding under Pilot's ribs when she was scanned, because we hadn't seen them.

When it was apparent that no more puppies were on the way, Lizzy and I breathed another huge sigh of relief. Pilot had delivered ten Golden Retriever puppies, all cream-coloured as we'd planned by the selection of her mate.

She'd had six boys and four girls in six hours and five minutes, in a labour that could have gone on for up to thirty hours. She could have got into difficulties, and needed an emergency trip to the vet: whelping doesn't always run so smoothly, and deciding on this course requires much careful thought and consideration. Pilot had come through the whelping just fine, and all of the puppies had a good birth weight, ranging from 394g (13.9oz) to 564g (20oz).

Lizzy had ordered a pack of 12 Velcro whelping collars especially made for newborn puppies to wear as a means of identification. And, just as well, too: with so many puppies all the same colour, we needed the collars to know which was which. And it was easy to adjust the collar size as the puppies grew.

So, ten puppies; all colour-coded, all named after cars.

Shaun, Lizzy's husband, a great car enthusiast, had chosen eight boy and eight girl car names in case the litter was all boys or all girls.

In birth order we had –

- Liana (pink)
- Bentley (blue)
- Dodge (red)
- Porsche (purple) Lily's original name, which I changed in honour of Lil, my mother-in-law, when Lily was five weeks old
- Spyker (green)
- Aston (black)
- Elise (cream) (she became 'Yellow Elise' within three weeks because her cream collar provided no contrast to her coat)
- Mini (orange)
- Morgan (brown)
- Royce (bright green) referred to as 'BG Royce' for there was already another green-collared puppy, Spyker)

I initially chose a purple collar for Lily for the alliteration of 'Purple Porsche.' As the symbolic colour of royalty, it was perfect for her when, at ten weeks old, she became our 'Princess Bunches,' her ears making it seem as though her hair was in bunches.

Purple was definitely Lily's colour. Purple combines blue and red: blue for calmness, reflected in her calm and calming temperament, and red for energy, her mental energy to live and her physical energy to survive.

I recorded a wealth of information about each puppy –

- birth order – first, second, third, and so on, by collar colour
- boy or girl
- ability to nurse
- name of puppy
- placenta delivered or not
- time between puppies
- presentation position – front feet first, hind feet first, or breech
- weight

Now that all the puppies had arrived and had been dried and polished, Lizzy and I gazed in wonder at Pilot with her ten babies nuzzling into her. It was a lovely sight, but not quite the dream we'd imagined. We had serious concerns, as Liana still hadn't grasped the essentials of nursing, and Lily was still covering herself and everything around her in milk.

My daughter and I looked at each other. Something wasn't right with two of the puppies. What we didn't know, then, was that we needed to be even more concerned about a third pup ...

A trio in trouble

Pilot was very good at letting her babies nurse, but keeping them clean was another matter, and it took her five days to grasp that aspect of motherhood, along with helping them urinate and defecate. Because there were so many puppies, Lizzy and I helped her with these demanding tasks.

I wish those had been our only concerns, but they weren't.

On Day 2 of Liana's life, we knew she was having real trouble with feeding. She clearly could not nurse, but at least a solution was at hand: literally, a solution ... for baby dog milk.

In preparation for an eventuality like this, I had bought baby dog milk powder, a complete milk replacer feed for puppies from birth to eight weeks old. I also had eyedroppers, bottles, teats, and sterilizer.

I mixed the substitute solution of baby dog milk powder, and gave Liana a few millilitres (ounces) from an eye dropper at six that evening, and the same again at nine. I had already made an appointment for the vet to see her the next day.

The next morning, I weighed Liana as I did every puppy, and discovered to my great sadness that she had lost weight, from 400g (14oz) when she was born to 350g (12.3oz).

Liana travelled to the vet's surgery in a ventilated cardboard box, with probably more than enough towels and a heat disc to keep her warm. I put Bentley in with her for even more warmth and company. Though Liana was too young to be able to hear or see him, I was sure she could sense him. Bentley was also there so that the vet could compare Liana to him, her clearly thriving litter-mate.

The vet found that Liana was badly dehydrated. Putting her finger in Liana's mouth, she pronounced, "She's got a very strong sucking reflex. She must just not be able to nurse."

I asked if another vet could see Liana, and she agreed, taking Liana out of the examining room to a colleague, who also could not explain why the puppy was so dehydrated.

Giving Liana four hydration injections, she told me I needed to feed Liana as much baby dog milk as I could. Every two hours, in fact, day and night. Could I manage that?

"Absolutely," was my response.

Life is beyond precious, and once it is gone, it is gone forever. If feeding Liana every two hours, day and night, was what it would take to keep her alive and give her a good life, then that is what would happen. I knew I could count on Lizzy, my daughter, who was as determined as I to help the little puppy.

The vet didn't want Liana fed by eyedropper anymore, and certainly not by bottle. She gave me two 1ml (0.04oz) syringes, and, feeling optimistic about how much Liana would take, also gave me two 2.5ml (0.09oz) syringes, with a 5ml (0.2oz) syringe thrown in for good measure.

She wanted to see Liana the next day to find out how the puppy was doing. Of course, she was impressed by Bentley: he was doing just fine.

By 11 that morning I was home with the puppies. I prepared some baby dog milk, and syringed it into Liana's mouth. Poor hungry little puppy. I got 2ml (0.07oz) into her using the very thin 1ml (0.04oz) syringe, refilling it once. Feeding her took a while. She couldn't manage much at a time without some of it coming out of her nose, and every time that happened, I cleared her nostrils using the aspirator, the 'blue squidgy thing,' as Lizzy sometimes called it. Over the next 11 hours, Liana took 16ml (0.6oz) of baby dog milk substitute, using and refilling the 1ml (0.04oz) syringe.

The next morning, Day 4, I took Liana back to the surgery. At our 10:20am appointment, I told the vet that we had fed Liana every two hours. Since midnight, Liana had had 18ml (0.6oz) of baby dog milk. Lizzy had done the first night shift, even though she had to go to work at 6:30 that morning; feeding Liana at midnight and again at 2am, before waking me at 3am and going home for a couple of hours' sleep.

The vet was pleased that Liana was still hydrated. Getting milk into her by syringe was working, and we needed to continue to feed her every two hours, day and night.

Liana hadn't gained any weight, but at least, she hadn't lost any more.

Had I been thinking more clearly, I would have asked about tube feeding Liana instead, which I could surely have learned how to do. But this option was not suggested, and, besides, I had another puppy on my mind now: Mini.

Mini was the smallest puppy in the litter, with a birth weight of 394g (13.9oz) that was perfectly normal. On Day 1, she had been nursing with enthusiasm, but now wasn't nursing at all. She had become listless, was floppy, and wasn't thriving.

So, that same morning of Day 4, Mini went to see the vet with Liana.

After examining Liana, the vet picked up Mini. The puppy lay in her hand unmoving. Agreeing that the puppy was listless, she said that, otherwise, she was okay. If I was still worried about her by the evening, I should bring her back.

At six that evening, Lizzy was sitting cuddling Mini, keeping her warm, making sure she felt cherished. Mini hadn't nursed at all that day, and, with every breath, she moaned softly.

Half an hour later I took Mini back to the surgery.

The vet held Mini in one hand. Mini was no longer making any sound, and her laboured breathing seemed shallow. She lay very still.

There was obviously something terribly wrong with Mini internally, but the vet was not prepared to operate on a four-day old puppy, as she was unsure of what she'd find, or whether anything could be done to help Mini.

I knew I was going to have to make a decision about Mini, but didn't want to make it alone. I asked the vet what *she* would do. With great reluctance, she told me, softly and slowly, "I'd let her go."

With more sadness than words can express, I signed the paper to have Mini put to sleep, only after the vet guaranteed that Mini wouldn't feel any pain in the process. The puppy had suffered enough in her four days of life.

The next day I wrapped Mini's little body in a small, soft towel, and placed her in a cardboard box with a letter I had written to her, telling her how much we loved her. Shaun dug her grave, and we buried her in the garden, under the statue of an angel.

Lily – one in a million!

Now there were nine puppies.

It was a busy time, looking after them. Thankfully, Pilot had become more maternal, and was taking much of the work off our hands.

We were syringe-feeding Liana every two hours, and I was making sure sterilized syringes, cooled boiled water, and prepared baby dog milk were always ready.

I weighed the puppies at least twice each day, too.

By the end of the first week, puppies should be double their birth weight, and Liana wasn't. Weighing 400g (14oz) at birth, and then dropping to 350g (12.3oz), she had gained very little in a week, her weight increasing to only 364g (12.8oz).

As the hours and days went by, my daughter and I got more baby dog milk into Liana, alternating syringe feeding with clearing the milk from her nose with the aspirator.

On Day 10, Liana was finally back up to her birth weight. Syringe feeding and aspirating were working, and she was growing and getting stronger. She even stood up and defecated by herself!

Lily hadn't doubled her birth weight, either (up from 406g (14.3oz) to 540g (19oz or 1.2lb)), but at least she hadn't lost weight as Liana had.

Lizzy spent a lot of her puppy time supervising Lily while I looked after the others. Lily nursed with the same enthusiasm she had shown on Day 1, but, just as she had that day, was taking in milk faster than she could swallow it. Her muzzle, chest, front legs, and the veterinary fleece pad she was on were constantly drenched with milk. There was lots of laundry to do to keep the puppies' bedding clean and dry.

Both Liana and Lily continued to gain weight, but I don't think either puppy enjoyed the feeds. I certainly didn't, as I was too concerned about them. Both always had milk coming out of their noses, and Lizzy and I become experts at using the aspirator.

It would have been lovely if we had had more time to enjoy the seven puppies who were thriving, but our concerns about Lily and Liana meant we were too busy.

When Pilot's puppies were two weeks old, their first dose of liquid wormer became due, and the wormer of choice was a 10 per cent liquid suspension given by syringe: 1ml (0.04oz) for every kilo (2.2lb) of body weight.

Worming the seven puppies who had no problems nursing was relatively straightforward. I simply held each one on my lap, and, with a syringe loaded with the appropriate amount of wormer, slowly depressed the plunger on the syringe until all of the wormer had dripped into the puppy's mouth, and was swallowed.

It was going to be a challenge with Lily and Liana to ensure that all of the wormer ended up in their tummies ...

I was particularly concerned about Lily. I was used to giving Liana liquid baby dog milk, which I'd never done with Lily: I didn't know how she would manage.

With Liana weighing only 550g (19.4oz, or 1.2lb), and Lily weighing 640g (22.9oz, or 1.4lb), the girls each needed just 0.5ml (0.02oz) of liquid wormer. Giving this to Liana was much different to giving her baby dog milk, for she had it drop-by-drop because I didn't want it to come out of her nose.

I took the same, slow approach with Lily, and managed to drip the wormer slowly enough into her mouth to ensure it didn't come out of her nose. Somehow I was successful with both girls.

The two-week dose was done. I had the five-week and eight-week doses to look forward to. I could wait, believe me, and I'm sure the puppies were happy to wait as well.

Day 19 was life-changing.

Liana's records show that Lizzy fed her at midnight and 2am. Milk came out of her nose.

I fed her at 4am. Milk came out of her nose. Poor puppy.

At her 6am feed, determined to make this a happier occasion for Liana, I set myself a really special challenge: I was absolutely determined that in this feed she wouldn't suffer the distress and discomfort of having milk come out of her nose.

I couldn't manage it. No matter what I tried, milk *still* came out of her nose.

Desperate to find out what was going on with her, I opened Liana's mouth. My heart sank. To my utter shock, I saw this dear little puppy had a cleft palate, that extended to almost halfway along the roof of her mouth.

No wonder she hadn't been able to latch onto Pilot that first day. And no wonder that she always had milk coming out of her nose.

With tears in my eyes as well as in my voice, I rang Lizzy and broke the bad news.

A few hours later, at the veterinary surgery, I signed the paper to have Liana put to sleep.

Knowing what I do now, I would not have made the same decision about Liana, and would have instead requested – and even insisted – on immediate referral to a specialist veterinary clinic, where she could have had her cleft palate evaluated for the best course of action. Tragically, I was not advised that there were other options, such as tube feeding and surgery.

Afterward, I brought her home and prepared her for burial, just as I had with Mini. With Shaun at work, I dug into the soil by Mini's grave, and buried Liana beside her.

Now the angel had two baby dogs to love and care for.

We were well and truly heartbroken, and so distraught that we'd put Liana through the difficulties of a feed every two hours for weeks, possibly distressing her. And for what?

I don't know why the vet hadn't seen Liana's cleft palate on Day 3. Perhaps it wasn't very evident because Liana had been so small. Or perhaps the vet hadn't much experience of newborn puppies. After all, not many newborn puppies end up at a veterinary surgery. Just as every doctor in general practice isn't a neonatal specialist, not every veterinarian is a specialist in newborn puppy care.

Even now, years later, I feel tremendously sad, thinking about little Liana.

I remembered what had happened with Maggie's litter of eight puppies. One puppy had died during the first night, only hours after coming into this world. Maggie's vet told me it is rare for an entire litter to survive, and that a quarter to a third of the puppies can die during the first week. Pilot had had ten puppies, which meant that up to three of them *could* die.

I'd already lost two puppies, and was determined that I would not lose a third! I did not realise, then, however, how soon that could have happened ...

Survival tactics

I never begrudged the hours I spent trying to help Liana stay alive, but once she was with Mini and the angel, I realized just how much time looking after her had taken.

Feeding Liana every two hours had been just the beginning. Each feed had taken at least 15 minutes, after which, Lizzy and I would gently stimulate Liana to urinate and move her bowels, which sometimes took a while. The 'while' always seemed a lot longer at three in the morning.

So much about giving Liana baby dog milk took time. Having cooled, boiled water at hand; mixing the powder with it in a sterilized container; heating the milk to the right temperature; pulling up the plunger on the syringe and filling it; cleaning the aspirator, and, afterwards, sterilizing everything for the next feed.

It seemed that no sooner had Lizzy and I finished one feed than it was time to feed Liana again. Lizzy and I were certainly kept busy, looking after the seven thriving puppies, as well as aspirating Liana as we fed her, aspirating and cleaning Lily as she nursed, changing and washing the milk-soaked and soiled veterinary fleece bedding, and looking after Pilot, and cleaning her milk-saturated side where Lily had nursed.

Once I had the time to devote to helping Lily survive, I appreciated how strong her will to live was, when she must have been constantly hungry, thirsty, and undernourished. It was a miracle she was still with us.

When Liana was put to sleep on Day 19, she weighed 648g (22.8oz or 1.4lb), and Lily weighed 706g (24.9oz or 1.5lb): not much more than Liana. The only other surviving girl puppy, Elise, weighed 1718g (60.6oz or 3.7lb). I was shocked to realize that Elise weighed over 1000g (35.3oz or 2.2lb) more than Lily! One thousand grams! Elise always had been heavier than Lily, weighing 430g (15oz) at birth to Lily's 406g (14.3oz), but 1000g (35.3oz or 2.2lb) was a whopping difference now.

It seemed to me that Lily was only just hanging onto life by her little nails, and slowly losing her grip. She couldn't go on as she was. I had to help her ... and fast.

The earliest that puppies should be weaned is at three weeks, on Day 21, when they are sufficiently developed internally to handle and digest solid food. Hoping Lily's stomach and gut would process solids enough to nourish her, I began weaning her two days earlier than recommended, on Day 19: the very day that Liana was put to sleep. The worst that could happen was something I didn't want to think about. It wasn't true that I had nothing to lose, and it was true that I had everything to gain: Lily's life.

I'd bought tins of puppy mousse for weaning any puppies who needed special attention. Clearly, Lily was the only one who needed the mousse, so I decided that the others would be weaned with soaked and mashed puppy starter kibble. I believed they'd be fine with it, and they were.

I started Lily on the puppy mousse that day at noon. At first, I held her on my lap with one hand, and, with the other, offered a tablespoon (10g/0.4oz) of puppy mousse for her to lick off a plate. She was so hungry that she tried to gobble it all down as quickly as possible, causing her to cough every three seconds or so, even when I slowed her by putting only a little puppy mousse on the plate. She also sneezed puppy mousse out of her nose several times. I used the aspirator to clear her nostrils.

Letting her lick puppy mousse from a plate on my lap clearly wasn't working, not when she was sneezing it out and clogging her sinuses with it. So, with the plate on the table beside me, I put my finger into the puppy mousse and let her suck a tiny amount from this.

That worked better. It took a while, but, eventually, bit-by-bit, and fingertip-by-fingertip, Lily sucked all of the remaining puppy mousse from my finger, and only a little came out of her nose. This really was progress. For the first time in her young life her tummy was as full with food as it could be, and Lily fell into a sleep of deep contentment instead of sheer exhaustion from her mainly futile efforts to fill her tummy with her mother's milk. Her hunger pangs had been banished, and I would make sure they never returned.

By Day 21, after just two days on puppy mousse, Lily had gained 60g (20oz) in weight, 8.5 per cent of what she had weighed the morning of Day 19. The photo of her on Day 21, asleep between Elise and Bentley, is startling, as it shows her as a third their size. At least, now, more food was going into her tummy.

Two days later, Day 23, became a 'Good Day/Bad Day.'

It was a good day ... thanks to the puppy mousse, Lily had finally doubled her birth weight, although it had taken her three weeks to do what her litter-mates had done in one. Now she weighed 812g (28.6oz or 1.8lb), a big improvement from 706g (24.9oz or 1.5lb) on Day 19.

It was a bad day, though ... because Lily needed to go to the vet. From just before I'd started weaning her, she had begun making dreadful, snorkelling sounds while she nursed and, then, when she licked puppy mousse.

This time, a different vet was on duty. After examining Lily, he said he thought she had a sinus infection, which I suspected may have been caused by the milk and puppy mousse in her sinuses. The vet gave her an antibiotic injection. He was also concerned about her weight issue, and the way she always had milk coming out of her nose when she nursed. He had already looked in her mouth but did so again, just to make sure she didn't have a cleft palate. The roof of her mouth was fine, and he simply didn't know why she was having problems nursing, though agreed I had been right to start her on puppy mousse.

Three days later, on Day 25, Lily returned to the same vet for a check-up. Lily's sinus infection had cleared, and the vet told me that she may have had an infection in her lungs as well from aspirating milk or puppy mousse, but that, too, had cleared, if so.

The vet commended me for looking after Lily as well as I had, with Lizzy's help, and then asked how I was going to look after her as she got bigger. How would I feed her when she couldn't eat by herself? Hydrate her when she couldn't lap water? Giving her tiny amounts of puppy mousse from the end of my finger, and water from a 1ml (0.04oz) syringe was keeping her alive at the moment, but how would I manage when she was a full-grown Golden Retriever: thirty times bigger than she was now, with food and water requirements to match?

Yes, he had a very good point. I hadn't thought ahead about what Lily

Lily - one in a million!

would need as an adult, as I was living not just day-to-day with her but meal-to-meal.

The vet then asked me whether I felt it was fair to Lily to keep her alive, given the real risk of recurrent bouts of pneumonia caused by aspirating food into her lungs, which was more a certainty than a possibility. She might have chronic sinus infections, too, giving her pain and involving years of expensive antibiotic treatment. He was obviously considering Lily's quality of life from both a professional and compassionate viewpoint, and assured me that he would support my decision to put Lily to sleep if that was what I wanted to do.

Lily was already hard work, which was only going to become even harder as she grew, and her needs grew with her.

And we still didn't even know what was wrong with her.

But if I was going to make the heartbreaking decision to have her euthanised, I simply couldn't decide there and then. I took Lily home.

That afternoon, a breeder rang to find out how I was managing with ten puppies. I told her what had happened to Mini and Liana, and mentioned the problems I was having with Lily, and what the vet had said about putting her to sleep. The breeder agreed, and advised I 'turn her into a little angel.'

I have great respect for this breeder, who had already given me lots of valuable help and very good advice when Pilot was in whelp. Her advice now to euthanise Lily made sense from her point of view because it would not be economically viable to do otherwise.

When 23 breeders, each having at least 20 years' experience of breeding dogs, were asked had they ever made superhuman efforts to save a puppy who appeared to be dying, more than half of them said no, and those who had made the effort regretted having done so because the puppy died in spite of this, leaving them distraught. One breeder reflected, "It kills me to lose one."

After speaking to the breeder that afternoon, I cuddled Lily. She was so small, so defenceless, so dependent on me as she fell asleep snuggled on my lap, I simply didn't want to think about what was in my mind.

So I didn't think about it until Jackie, the breeder I'd bought Pilot from, phoned that evening.

Jackie congratulated me on bringing Lily along as well as I had, and encouraged me to "keep up the good work. Cherish every day with Lily. You don't know how long she will be with you, but however long it is, you'll always regret it if you don't do your best for her."

So the vet and a breeder believed Lily should be put to sleep, while my friend, Jackie, also a breeder, and Lizzy and I believed she should be helped to survive ... but only if she could have a life worth living.

Even if it hadn't been three against two, my decision was an easy one: I was going to help Lily not only to survive but survive to live life to the full!

Within days of my decision, it was time to again worm Lily and the other puppies, at five weeks old and ready for their second dose.

Strangely, whilst solids were a big problem for Lily, liquids were an even bigger one. And the wormer of choice would just have to be a liquid, wouldn't it?!

By now, Lily was eating almost 200g (7oz) of puppy mousse a day, and had gained weight. At 1450g (50.4oz or 3.2lb), she needed 1.5ml (0.05oz) of wormer, which may not sound like much, but it looked like an enormous amount to me after

I loaded the syringe, and began dripping the wormer into her mouth, ever-so slowly so it wouldn't go into her nose and sinuses.

I knew that Lily couldn't live on puppy mousse forever. Her litter-mates were thriving on puppy starter kibble, softened overnight in water, then mashed lightly with a fork. They happily and messily ate this in preparation for going to their new homes.

I tried giving Lily softened and mashed puppy starter kibble by half-teaspoons on a plate, but the results weren't good.

My notes show –

- Day 50: Quite a bit of choking
- Day 51: Gagging again and coughing up bits of kibble
- Day 53: Much gagging, and coughed up bits of kibble
- Day 55: Gagged and coughed up kibble. Didn't finish what she had

On Day 56, I used a small food processor to turn the soaked kibble into the same consistency as puppy mousse. I smeared the pulverized food onto a plate, a very little at a time, for her to lick. The result was 'Great!' and 'Perfect!' according to my notes.

From that day, she scored an 'A+' at each of her five feeds, progressing from 65g (2.3oz) at each meal to 70g (2.5oz), 80g (2.8oz), 90g (3.1oz), and 100g (3.5oz).

At eight weeks old, Lily and her litter-mates required their third dose of wormer. Lily weighed 3700g (130oz or 8lb); the only puppy I could still weigh by putting her in an empty shoebox on my kitchen scale, which went up to 5000g (176oz or 11lb).

While I was thrilled with Lily's weight gain, the downside of it was that she needed almost 4ml (0.2oz) of wormer. Oh, joy! That was a lot of wormer, almost filling my 5ml (0.2oz) syringe. It took me 30 minutes, multiple attempts, and serious perseverance to get it into her tummy, and not in her nose or sinuses, on her paws, her chest, her muzzle – or me.

Lily and I were coping pretty well with feeding. She was no longer licking my finger but rather small amounts of processed soaked kibble smeared like a paste over the bottom of a bowl. I held the bowl at an angle so that she didn't have to bend down to her food. She still gagged and choked while she licked the kibble paste, though.

When she was about ten weeks old, I tried yet another way of feeding her. I abandoned the small food processor and its soaked kibble, and, instead, hand-fed her unprocessed soaked kibble, one piece at a time. I filled a syringe with the liquid I drained from the kibble and squirted the liquid into the palm of my hand, which she then licked up.

It took ten minutes to feed her a meal, during which I had to listen to her swallow, and watch her nose for discharge. If this was clear and she was swallowing okay, I gave her another soaked biscuit. Hand-feeding her was working fairly well, although no meal was without difficulties.

At ten weeks, Lily had her first vaccination, two weeks later than the other puppies because of her slow development.

At twelve weeks, the vet gave Lily her second vaccination injection, and asked how I was managing with her.

Lily – one in a million!

I told him Lily was still gagging and choking at each feed (the 'Great,' 'Perfect' and 'A+' scores hadn't continued).

How was I feeding her? he asked.

I told him about how I was hand-feeding Lily, and also about how her tummy often swelled so much, as though she was full of air, rather than food. (She actually was full of air, not food, although I didn't know it at the time.)

Subtle clues in the vet's body language and facial expression told me he was more alarmed by what I told him than he let on, but I was becoming alarmed, too, and wondering why she was having so much trouble.

Two weeks later, at the vet's suggestion, I took Lily back to him so that he could look deep down her throat whilst she was sedated. At the end of the examination, he admitted he couldn't see any reason for her trouble. There was definitely no cleft palate. His notes show that, astutely, he suspected pharyngeal dysfunction, which meant nothing to me at the time.

He strongly recommended that Lily be seen by a specialist vet.

How much would it cost?

A lot. And Lily had no pet insurance, and would not be eligible, in any case, as this was a pre-existing condition.

The vet's notes recorded that "... puppy not doing very well but owner will soldier on because referral would be too expensive."

The next morning, my son, Marcus, phoned me, as he still does every weekday morning, and asked how Lily's examination had gone.

I told him that the vet still didn't know what was wrong with her, and wanted to refer her to a specialist veterinary clinic, but I couldn't do that as it would cost so much.

Marcus asked, "What if whatever is wrong with Lily can be fixed by an operation? How would you feel, knowing she could have been helped? Don't you remember what Grandma said? 'Money can always be replaced but an opportunity can't!' Here's the opportunity. Find out what's wrong with her."

He was right. I did need to know what was wrong with Lily, especially if she could be cured.

I realize now how desperately – but unrealistically – I'd been hoping Lily would simply outgrow her feeding problem, or overcome it. I'd read about 'gluttony syndrome' which, it seems, a few puppies develop when they begin eating solid food.

They make snorkelling sounds as they eat, have food coming out of their nose, and may blow bubbles through their nose into their food and water. Soon after eating, the puppies begin to gag, choke, and run around in circles, trying to leave behind what is affecting them. They may even have so much trouble breathing for a few minutes that their tongues turn blue. Apparently, though, they begin eating properly within a week of weaning.

I would have been frantic, and, doubtless, Lily would have been, too, had she been gasping for breath so desperately that her tongue turned blue. But with that exception, and the last bit about beginning to eat properly, Lily's symptoms seemed a lot like those resulting from this syndrome, although, she hadn't fed well for the first month or so until weaning began.

Gluttony syndrome was mentioned in only one paragraph in a book a friend had loaned me, but I read that paragraph so many times. I really wanted this to be what Lily had, because it meant she would outgrow it soon after weaning. But she was three months old, now, and still had a serious problem. Lily hadn't been able to nurse successfully from the moment she'd been born, and still couldn't eat properly; couldn't even lap water. I had to admit what I knew in my heart: whatever was wrong with her, it wasn't gluttony syndrome.

The morning of my telephone conversation with Marcus, I phoned the surgery and asked the vet to refer her, and he wasted no time in doing so. (A gifted veterinarian, he once told me that if I could keep Lily alive, she would grow to be almost as big as a normal-sized female Golden Retriever, and he was right. To my regret, he retired soon after he referred Lily to the specialist clinic.)

No one could have foreseen how highly unusual the outcome would be.

Visit Hubble and Hattie on the web: www.hubbleandhattie.com
hubbleandhattie.blogspot.co.uk
• Details of all books • Special offers • Newsletter • New book news

A diagnosis

Within a week of Lily's referral to a specialist veterinary clinic, I received an appointment for an hour-long consultation and assessment.

However, before Lily could set one paw through the door of the clinic, she had to be inoculated against kennel cough, and had this when she was 14 weeks old.

Three weeks later, when Lily was 17 weeks old, the European Specialist in Small Animal Surgery (Soft Tissue) saw her. The specialist met us in the reception area of the clinic on the appointed Thursday, and, as we walked to her office, remarked that she had seen several dogs with a cleft palate just in the previous three days.

I'd had no idea that operations on cleft palates were done on dogs at all, never mind almost routinely. If I had known this, I would have had Liana operated on, but no one had mentioned it to me.

A vet whom Lizzy – desperate to know what was wrong with Lily so that she could help her – had consulted online without my knowledge soon after Liana was euthanised, had diagnosed Lily's problem as a cleft palate, but only because he had absolutely no idea what else could be causing Lily to choke and have milk come out of her nose.

As I talked through Lily's history and the specialist examined Lily, I was enormously impressed by her knowledge, the examination she gave Lily, and the assessment. After a good forty minutes, the specialist sat down at her desk and told me, "Well, she certainly doesn't have a cleft palate. I think she may have pharyngeal achalasia."

"Yes, that's what I was thinking," I said. Or would have – maybe – if I'd thought of it in time. Such an obviously laughable untruth coming from me, while nodding slowly and wisely, my brows furrowed, would have helped relieve the tension of my concern about Lily at that moment. Of course, I'd never heard of this condition: I couldn't even pronounce it! Not then, anyway.

I was still getting my head around this many-syllabled suspected diagnosis, when she expanded further: "Actually, cricopharyngeal achalasia." Another two syllables!

Then she drew a diagram to show me what this involved, and explained everything very clearly.

Apparently, Lily presented with all the symptoms of this diagnosis –

- frequent swallowing, even when there was no food to swallow
- regurgitation
- gagging
- nasal discharge
- difficulty swallowing food
- even greater difficulty swallowing liquid

Lily's seemed a textbook case. In fact, a textbook case was the only case the specialist had encountered: in her 26 years of being a vet, she had never seen a dog with cricopharyngeal achalasia. I was enormously impressed she could even say the name of the condition, much less know about it, given she'd never seen it before.

She told me there was a surgical procedure that might correct it: myectomy of the cricopharyngeal muscle. In language I could understand, she explained that this involved cutting or removing all or part of the muscle at the top of Lily's oesophagus. The specialist seemed reluctant to try this option, however, and told me there was a another surgical procedure that could benefit Lily: fitting a gastronomy tube through her side so that she could be fed directly into her stomach.

No more choking, no more coughing, no aspiration pneumonia, because nothing would go from her throat into her lungs, and she'd always get the right amount of food.

Nevertheless, if I could prevent – or even delay – Lily having a stomach tube, I wanted to, as I wasn't keen for her to have one.[1] She was four months old, a happy, active puppy. With a tube in her side, I would worry – rightly or wrongly – that it would be disturbed during her roughhouse puppy play with Pilot and Bentley as they raced around, full of the joys of life. I didn't want to confine her to the house because of my concern about the tube, and see her press her nose against the French windows, watching her mother and brother chase around in the garden without her. I was worried it might affect her natural life, and so I declined the tube.

The specialist understood my reasons, and stressed she wanted to obtain a firm diagnosis about what was the matter with Lily before any treatment, in any case. She needed to do some tests on Lily, and said she'd schedule her for a barium sulphate fluoroscopy for the next day, a Friday.

A fluoroscopy? Really? I thought a fluoroscopy was a diagnostic procedure used only on people! I was astonished that veterinary science meant it could be used to diagnose what was wrong with Lily. It was incredible that the specialist and her team would be able to watch exactly what was happening with Lily's swallowing reflex.

I took Lily back to the veterinary clinic the next morning for the fluoroscopy, along with a container of her usual, water-soaked food. It was quite a wrench to leave her there, but I knew she was in completely safe hands.

By early afternoon, she was ready to come home.

In a consulting room, a nurse explained how they'd been able to watch exactly what was happening as Lily swallowed, or attempted to swallow, her food. Lily had managed some soaked kibble the consistency of slightly runny pâté when it was put in a dish. She certainly had difficulty swallowing it, but experienced a lot more difficulty with water.

Fortunately, Lily's cranial nerves were normal, and her gag reflex was tight,

---

[1] Two years after Lily's diagnosis, my neighbour, Hilton, said he'd refused a gastronomy tube for his Boxer, Fenster, who had megaesophagus. With the risk of recurrent pneumonia from aspirating food, Fenster was expected to suffer recurrent, debilitating pneumonia by aspirating food into his lungs. His life expectancy was 20 months. Hilton's intensive management enabled Fenster to thrive for ten more years. He died from a suspected heart attack.

which was especially good news because this would help her to cough up food and water before it went into her lungs.

The nurse told me that the specialist would phone me the following Monday to discuss in detail the results of all the tests.

The weekend gave me time to practice saying cricopharyngeal achalasia, and Google the condition. What I found online didn't make for happy reading.

Swallowing is a complex process requiring the tongue, hard and soft palates, pharyngeal muscles, oesophagus, and gastroesophageal junction to work together. More dogs have swallowing problems than you'd think, especially Golden Retrievers (Lily's breed, of course! Wouldn't you know it!), and Cocker Spaniels and Springer Spaniels. Other breeds predisposed to congenital swallowing problems are the English Bulldog, Chinese Shar Pei, Bouvier des Flandres, Boxer, and German Shepherd. Large breed dogs can also experience chewing problems.

I never would have thought that dogs would have problems with swallowing. I wondered how many were actually born with Lily's condition, and suspected it would be impossible to know, simply because dogs who had it probably died before a diagnosis could be made, let alone the condition managed.

Every study I read about cricopharyngeal achalasia emphasized how rare this condition is as a cause of canine dysphagia (difficulty or discomfort in swallowing). Learning that dogs presenting with it could end up as respiratory emergencies requiring intensive support didn't fill me with confidence about Lily's long-term prospects. Reading that the condition could be surgically treated had made me feel a bit better, but only until I read about the results of pharyngeal surgery.

One study of surgical treatment showed a success rate of only 65 per cent. That's almost a case of 'It might work, it might not' for an operation that benefited a little over half the dogs who underwent it.

Even worse, another study showed that the operation had a desperately high failure rate. Over a period of 12 years, 14 dogs were surgically treated for cricopharyngeal dysphagia, and, of those, eight had to be euthanised, either because of persistent dysphagia, or because of aspiration pneumonia after food went into their lungs. Of the six remaining dogs, three did not improve, and the remaining three grew worse.

No wonder the specialist hadn't seemed enthusiastic about pharangeal surgery for Lily ...

I also read that the condition wasn't usually evident until weaning onto solid food begins.

Wait a minute! What had I just read? The condition wasn't usually evident until weaning onto solid food begins?

Did that mean, then, that dogs with cricopharyngeal achalasia were actually able to nurse, to swallow milk – a liquid – so successfully that it went into their tummies for at least the first 21 days, because Lily hadn't been able to do that for one day, let alone 21!

Was Lily an extreme case, given that she'd had serious trouble nursing from the moment she was born? Or was it that she *didn't* have cricopharyngeal achalasia after all?

Whatever the answer, I doubted that Lily would have lived long enough to be weaned if we hadn't taken special care of her, alternately aspirating her and mopping up her and her surrounds of all the milk that hadn't gone where it should have. She could have drowned if Pilot's milk had flooded her lungs. As it was, she may have had the beginnings of aspiration pneumonia when she was treated for a sinus infection.

By Monday, in preparation for the specialist's call, I felt up-close and personal with cricopharyngeal achalasia. I could pronounce it as easily as the specialist had. I was ready for her call.

True to her word, the specialist contacted me on Monday after she had considered the results of the tests, and discussed Lily's very interesting and challenging case with colleagues.

Then she said, "Lily doesn't have cricopharyngeal achalasia: she has cricopharyngeal asynchrony."

The term 'achalasia' encompasses 'asynchrony,' but 'asynchrony' is an even rarer genetic disease. How much rarer? Enough that Googling it doesn't even come up with it in a bold print heading as 'achalasia' does. 'Asynchrony' is in small print – when you can find it at all.

So, as I understand Lily's condition, there is asynchrony (lack of working together at the same time) between contraction of the pharynx (the membrane-lined cavity behind the mouth and nose, connecting to the oesophagus) during swallowing and relaxation of the sphincter (the muscle that opens and closes the oesophagus).

I was fast going into information overload. It was enough for me to appreciate that swallowing is complex. The muscles involved in the procedure need to be synchronized to relax and contract at the right time. Lily's aren't synchronized, which means they don't relax and contract when they should.

The specialist told me that the barium sulphate fluoroscopy showed Lily's upper oesophagal sphincter muscle sometimes let the bolus (a soft ball of chewed food, in this case, Lily's soaked kibble) go down her oesophagus. Sometimes it shut and kept the bolus in her mouth, which prompted her to keep trying to swallow, and other times the muscle slammed shut with half the bolus left on one side (in her mouth) and the other half the other side (in her oesophagus). No wonder she found swallowing difficult, and often kept trying to do so.

I was told that whatever I fed Lily must be given at knee height, as she wasn't to stretch her neck to a bowl on the floor to eat. And her daily total amount of food had to be divided into three separate meals so that she wouldn't try to eat too much at one meal, and choke.

I certainly wouldn't have given Lily fewer than three meals a day because her meals were her main way of getting water. She couldn't have gone all day without what was, in essence, a drink of water.

The specialist commended me on bringing Lily along as far as my daughter and I had, and wished us continued success. If I ever changed my mind about the stomach tube, though, I just needed to let her know.

Then she added that everyone at the clinic who'd seen Lily, from herself to the receptionists, had found her a complete delight, an extremely pretty puppy with a wonderfully friendly temperament.

The specialist finished by saying: "Lily certainly ticks the 'Unusual' box!"

Having Lily diagnosed was absolutely worth every penny, as now I know what's wrong with her. It makes us feel sad for her, though, because, as Lizzy

Lily - one in a million!

reflected, even if we won big on the lottery, no amount of money could fix her.

People are definitely interested in Lily. On several occasions, I've taken her with me as a kind of 'Show and Tell' to give talks about her. People ask 'How do you feed her? How do you keep her alive when she can't eat; can't even lap water?' The answer to these questions is 'Intensive management by the owner,' which is what's on her medical record.

Knowing what is wrong with Lily has been enormously helpful, as I now have an idea of what is going on when she starts her repetitive and unproductive swallowing, air-gulping, and choking. If the day comes when Lily's difficulty with eating is so extreme that she keeps getting aspiration pneumonia, I will seriously consider a stomach tube for her. I've Googled it, and seen photos of more dogs than I could believe who have feeding tubes inserted through their side, and they seem happy enough. Dogs can even make their own fashion statement by wearing stylish covers around where the tube enters their side. But I still would rather not have one for Lily.

Meanwhile? I had to help Lily survive somehow so that she could have a great life, and that meant feeding her incredibly carefully and listening to her every swallow, not only day-by-day but meal-by-meal.

I had no idea how creatively I would need to think to achieve this, how 'eureka' moments would occur when I needed them most, and how Lily and I would learn from each other vital lessons for life.

Lily, just minutes old. In birth order, right to left, Pink Liana, Blue Bentley, Red Dodge and Purple Porsche (Lily's original name).

A miracle of survival. Day 21, and Lily (middle) was a third the size of litter-mates Elise (yellow collar) and Bentley.

A tired Pilot and her ten puppies.

The angel watches over the puppies who died: Mini (4 days old), and Liana (19 days old).

Lily playing alone, at four weeks old. Weighing 1250g (44oz/2.7lb), she was less than half the average weight of her litter-mates 3087g (108oz/6.8lb).

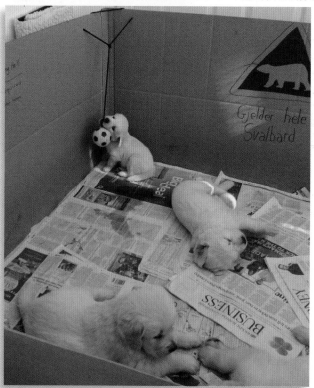

Lily – so small; so vulnerable – resting against Shaun's leg.

Lizzy arranged the eight puppies in a rainbow of collar colours, although Lily wouldn't stay still!

Left: Lily, five weeks old, in an improvised 'doggie carrier' ...

Left: ... and at 6 weeks old, in her own puppy carrier, held safely and snugly by an interior harness ...

... now hard to believe that Lily, almost three years old in this picture, ever fitted into this puppy carrier!

Lily at seven weeks beside a 95g (3.5oz) jar of coffee, 13cm (5in) high. She was now half as big as her litter-mates.

Left: Lily – aka 'Princess Bunches,' at ten weeks. Her ears make it seem as though she is wearing hair bunches.

Lily at ten weeks, after her first vaccination, with Pilot on the patio.

Lily, 16 weeks old,
snuggling with Pilot.

Below: Lily, almost 14 weeks old, with Pilot.
Mother and daughter: so alike.

Lily, at almost 12 weeks
old, with one of her
lollies – about the size
of a fifty pence piece –
which provides her with
30ml (1oz) water.

Lily, so very pretty at a year
and seven weeks old.

Above: Lily, Pilot and Bentley, after a fabulously muddy time. The puppies are clearly eager for more ... not sure Pilot is.

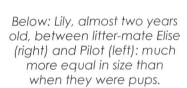

Lily, now almost a year old, watches Pilot enjoy the sea, but stays safely on the beach, almost as if she knows how dangerous it might be for her if she went in.

Below: Lily, almost two years old, between litter-mate Elise (right) and Pilot (left): much more equal in size than when they were pups.

Lily, aged two, focusing on Joshua reading to her in Pets As Therapy's Read2Dogs programme.

Lily gives Poppy an encouraging lick in a Pets As Therapy's Read2Dogs session.

The get well card that children in the Read2Dogs programme made, and posted to Lily after she was spayed.

Lily and I with Vanessa Duggan, winner of The Care Home Activities Coordinator Award 2016 for her work at Hamble Heights.

Get Well soon!

For a very lovely dog

Lily Is Lovely Young kind dog!

she has lots of friends.

We all hope you get better

Lily with Patricia Charles at Avon Park Residential Care Home ...

... and with Derek Coombes ...

... and with Elizabeth Wymer, all of whom loved spending time with her.

Left to right: Lily and Bentley, two-and-a-half-years-old, with Pilot at the beach.

... and at the same age, Lily enjoying her glowing Christmas present.

Lily with her marvellous vet, Dr Lucy Atkinson.

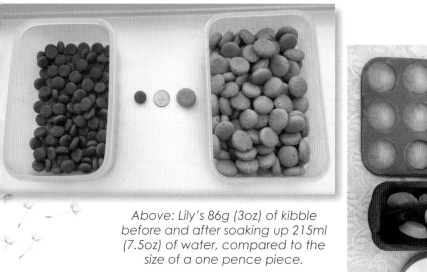

Above: Lily's 86g (3oz) of kibble before and after soaking up 215ml (7.5oz) of water, compared to the size of a one pence piece.

Right: Twelve frozen lollies in the tart tray, waiting to join the six already made.

Butter wouldn't melt ... Left to right: Bentley, Lily and Pilot waiting for a treat.

Waiting for her call-up from United ...

Above, right: There's nothing in Lily's rule book that says a girl can't carry the ball in her mouth!

Lizzy, Shaun and Bentley.

Almost three years old, Lily (right) and Pilot pose with me for a lovely shot.

Lessons for life

It's really heartening to know that, generally, we will do anything to help our dogs to not only survive, but also give them a wonderful life.[1]

Water

It's possible to live without food for a while, but not without water, which is why my decision to help Lily to live meant I had to find a way of getting water into her, given that she can't lap water. Figuring out how to do that has been challenging.

When she was ten weeks old (two weeks older than her litter-mates because of her compromised development), and went for her first injections, the vet asked how I was giving her water. I told him I was squirting amounts of 0.5ml (0.02oz) of water, barely enough to measure, from a syringe into the palm of my hand, and letting her lick this up.

My inspiration had come when I'd had a eureka moment from seeing deer at dawn outside my house. They were licking dew from the grass. If deer could imbibe some water by licking grass, couldn't Lily do the same by licking it from the palm of my hand? Admittedly, getting water into her that way was laborious and slow, but she was small, and didn't need much water.

The vet asked how I was going to ensure she drank enough water when she grew bigger – if she lived long enough to grow bigger.

Good question. I'd been living moment-to-moment with Lily, and hadn't thought years ahead. She weighed 5.1kg (11.3lb) that day. How would I get her to drink enough water when she was an adult weighing five times as much, perhaps 25kg (55lb) to 30kg (66lb)? It obviously wouldn't be by licking water from the palm of my hand.

The vet reiterated that he would support me if I wanted Lily put to sleep because of the problem of getting water into her. But I couldn't do that.

By the time Lily was four months old, I had begun getting extra water into her by giving her crushed ice cubes, which she licked from a small ceramic bowl.

Then I had another eureka moment! Make the ice cubes tasty. Give her ice lollies! Doggie ice lollies, minus the sticks.

I began draining the liquid from Lily's soaked kibble, and freezing it in tart moulds about 6cm (2.4in) in diameter. It's a method I still use for her now. Each ice lolly 'tart' has 30ml (1oz) of kibble-infused water in it. With the fragrance and flavour of her kibble, Lily really enjoys them.

---

[1] A fellow PAT volunteer, Jennifer, had a puppy, Sophie, the runt of the litter, who was born blind. Although four vets, one an eye specialist, suggested euthanasia, Jennifer chose instead to do her best to give Sophie a fantastic life. Jennifer took Sophie on walks, even sometimes off-lead, with Sophie circling Jennifer by listening to Jennifer's voice. Sophie died aged seven, having enjoyed the good life that Jennifer helped her to have.

Lily - one in a million!

After taking the lolly out of the freezer, I give it a splash of cold water, because I was born and grew up in Toronto, Ontario, Canada. What does that have to do with Lily's lollies? Quite a lot, actually.

There were fire hydrants on the streets of Toronto, and every winter we children were warned about what would happen if we touched our tongue to a frozen fire hydrant: it would stick to the fire hydrant and we'd be there forever. Or at least until spring. If we got free from the hydrant before spring, it would only be because we'd torn our tongue out of our mouth to get away, and had left it on the fire hydrant. The prospect scared us silly. Not surprisingly, we never stuck our tongue onto a fire hydrant even in the summer, just in case.

So that's why I always wet Lily's lollies: I didn't want her to get her tongue stuck to a frozen lolly just in case the horror stories I'd been told about tongues sticking to frozen fire hydrants really were true!

The lollies have made it possible for me to take her to the beach, though she's not allowed to swim because of the danger she might aspirate seawater. Taking a bowl of water to the beach for Lily would not work for her as it does for Pilot. Instead, I take three or four frozen lollies in a freezer bag with gel ice packs around them. I also take a collapsible plastic bowl from which she can lick her lollies when I hold it on my knees so that she needn't bend her neck.

Lily always finishes her meals with two lollies: they're her dessert. She enjoys them so much that she has more than a dozen a day. When she is thirsty after a walk or energetic play, she stands by her lolly bowl, and if I'm not quick enough to notice this, she will dip her tongue into Pilot's water bowl. That always gets my attention because I think "Wow! Can she lap water now?" But, of course, she can't, and doesn't even try. She knows lapping water will make her choke.

If she is still thirsty after finishing one or two lollies, she stays by the bowl until I give her more. I used to give Lily one lolly at a time, but every time Lily had a lolly, Pilot got a treat. So now I always give Lily two lollies at once to save having to take another out of the freezer, moisten it, put it into her bowl, and give Pilot another treat.

Most of the time Lily finishes the lolly without coughing or gulping air, but not if she gets distracted. If she hears someone at the door, for example, she'll leave the lolly while she charges off with Pilot to investigate. If she's gone long enough for the lolly to melt even a little, the water around it makes her cough and choke. Again, who would have thought liquid would be a greater problem for her than solids?

The main way of getting water into Lily isn't via the lollies, however, but her kibble: 1g (0.04oz) of dry kibble absorbs 2.5ml (0.09oz) of water. At each of Lily's three daily meals, she has 86g (3oz) of dry kibble soaked in water overnight. That gives her 215ml (7.5oz) of water at each meal.

A dog requires a daily minimum of 30ml (1oz) water per kilo of body weight. At 23.5kg (52lb), Lily's minimum daily water requirement is 705ml (25oz). The water in her three meals gives her 645ml (23oz), and adding to that the 360ml (12.7oz) of water that her twelve lollies provide, means she has about 1000ml (35oz) of water each day. No wonder Lizzy says Lily wees for England! I wouldn't have managed if I'd still just been using a syringe to squirt 0.5ml (0.02oz) of water into the palm of my hand for her to lick as I had when she was ten weeks old.

When Lily was about a year old, if she needed water more quickly than she could get it by licking a lolly, or if she began gulping air while licking a lolly, I used a manual ice crusher to grind her lollies into small bits to feed them to her by hand. I even took the ice crusher to the beach.

Now that she's older, when she needs the water from a lolly quickly, I break it into six or seven pieces using the handle end of an knife. She enjoys crunching the lumps. Lily's lollies are so essential that they cost Lizzy and Shaun the price of a new freezer. When Lily was 17 months old, they, along with their dog, Bentley, moved in with me for four months whilst their new-build was finished. Soon, Lizzy and Shaun realized they needed a separate freezer for their own food, as one of the drawers of my under-counter freezer is still used only for Lily's lollies.

Solids

Lily graduated from puppy mousse to soaked Golden Retriever junior kibble, which was ideal because she could manage a piece of kibble about as big as the nail on my index finger. The kibble supplied the nutrients for healthy joints, supported her immune system, and made her coat glossy.

The information on the bag indicated that she should have this until she was 15 months old, but I kept her on it until she was 18 months as it worked, and I was afraid of changing to something different.

When I eventually switched her to soaked adult Golden Retriever food, I had a big problem, as Lily couldn't manage to swallow this without significant choking and gagging. The pieces of kibble were just too big for her, and, even torn in half, were too dense for her to manage.

I was about to go back to junior food, but just in time I asked a canine nutritionist for advice, and I learned why junior food has an age restriction. It contains all of the calcium needed for growing bones and teeth, nerve impulse transmission, muscle contraction and heart rhythm regulation, but a dog of 15 months and above can be harmed by excess calcium, developing conditions such as hypercalcemia, with resultant increased urination, kidney damage, bladder stones, vomiting, thirst, and even coma.

The nutritionist recommended adult food; not for Golden Retrievers but Springer Spaniels, a medium-sized breed smaller than a Golden Retriever. Now fully grown, Lily is small for a Golden Retriever, but weighs the same as an adult Springer Spaniel – about 23.5kg (52lb).

The new kibble worked well from the first time I tried it with her, and we've stayed with it for her three daily meals and for treats.

Lily can't manage too much food at each meal, as this will mean she must swallow too many times. Having two soaked biscuits per mouthful, each meal entails about 60 productive swallows, and her daily diet comprises –

● 86g (3oz) of kibble in each of three containers, soaked overnight in the fridge for breakfast, lunch and dinner
● about 120 individual pieces of kibble (If my kitchen scales ever broke, I'd know how many pieces to place in each bowl.)

I must prepare all three of Lily's daily meals the evening before so that the kibble has time to soak up the water overnight. Have I ever forgotten to do that? Yes, once. Poor Lily, it was quite a shock the morning I saw her three clean and empty kibble containers on the kitchen counter: no soaked kibble. What was I going to give Lily for breakfast? I had to think quickly and creatively.

Before I took her and Pilot for their early morning walk, I measured the kibble

Lily – one in a million!

for her lunch and dinner, added water and put them in the fridge, hoping that the five or six hours until lunch would be enough for them to triple in size. I also put half of her breakfast kibble in water for it to soak up enough fluid by mid-morning to help hydrate Lily. The other half of her breakfast kibble I put in water so that it could absorb as much as possible while we were out on that first walk of the day.

By the time we came home, 40 minutes later, the kibble hadn't soaked up much water at all, but I fed it to her anyway: better half of her breakfast than none. With the kibble almost waterless, Lily had less trouble than usual with it. Afterwards, she had three lollies, so, at 30ml (1oz) per lolly, got at least 90ml (3oz) of water.

Lily had the other half of her breakfast for elevenses that morning, and, at one o'clock, had a fashionably late lunch, the kibble thoroughly soaked.

I take each meal out of the fridge as needed, and microwave the container of soaked food for 70 seconds to take the chill off it, then drain the liquid from it to freeze for her lollies. If she already has a dozen lollies in the freezer, I moisten Pilot's food with the excess liquid from Lily's food. Any excess liquid stays in a glass in the fridge until I need it for more lollies.

Up until Lily was almost three years old, it was rare for her to get through a meal in one attempt, even when I gave her two soaked biscuits of kibble at a time, listened to her swallow, and waited four or five seconds, if all was well, before giving her the next mouthful.

A straightforward meal with about 60 swallows – no drama, no interruptions – would take only about six minutes.

As she ate, Lily would nudge the bowl with her nose, sometimes twice if she thought I was too slow giving her the next mouthful. Her nudges are still sometimes so enthusiastic that if I don't have a firm hold on the bowl, pieces of kibble can be thrown out onto the floor. When that happens, if Lily lowers her head to eat them, I know she is okay. If she doesn't, I know she isn't okay, and there's something stuck in her oesophagus and she needs an immediate break.

Reading her subtle and not-so-subtle body language, listening to her swallow and hearing noises from her throat, I know when to tell her "Take a break. Go have a cough and come back when you can swallow." When necessary, she does exactly that. 'Cough, cough!' Maybe that's why she's never had aspiration pneumonia.

The fact that she will leave her food and wander around until she coughs to clear her throat shows she understands and trusts me completely, knowing I will be waiting with the rest of her food when she returns. It's frightening when she won't take any more food from me, because it means she can't. Food is stuck somewhere. When this happens, she wanders around distractedly, head down and tail between her legs, always near me; asking me to help.

At one breakfast, she did this. I patted her sides and back energetically, quickly and repeatedly for about ten minutes before she finally coughed out three pieces of soaked kibble that had become stuck somewhere between her mouth and her stomach. Scary.

I can't hurry Lily's meals, either. As my friend, Shauna, in Ontario always says, "The hurrier I go, the behinder I get." And that's exactly what happens when I try to make Lily's meals go more quickly. Feeding her ends up taking longer because too much food too fast makes her choke, cough, and keep swallowing repeatedly.

With all my concerns about Lily's meal times, it was a huge relief one Saturday in January when I had another eureka moment about how best to feed her. Because that particular Saturday morning's forecast was for temperatures of below zero, I wanted to know how cold it was before getting ready to take out the girls. Still in my dressing gown, I saw that it was -3.8°C (25°F) outside.

Did the girls – or I – really want to go out into the cold and the ice? No. They took their comfort break in the garden, and then I fed them.

I warmed Lily's food for 70 seconds in the microwave, drained it, and sat down with it. So far, so normal. As I was still wearing my long dressing gown instead of my jeans, I wrapped it closely around my legs to keep warm. Standing in front of my knees, my dressing gown wouldn't let Lily come as close to me as she usually did when I was wearing jeans. With her head reaching only slightly above my knees, she had to stretch her neck to nibble the food I was holding on my lap.

As usual, the first swallows were pretty well silent, so I didn't find it too remarkable that we were meeting the 'Triple S Challenge' of the single, silent swallow. She often got to swallow 17 before they became loud, repetitious, and unproductive. But now, as I continued to give her the next mouthful and the next and the next, I was listening with some surprise. Swallow 17 was silent and single. So was swallow 18. And 19. And 20.

I began to think I was on to something. And I was!

With swallow 60, into Lily's tummy went almost silently the last two pieces of water-soaked breakfast kibble. Incredibly, Lily had actually eaten a meal with no significant problems.

I could hardly believe it, and could barely wait to give her lunch to see if I could replicate breakfast's success by making her stretch her neck horizontally to reach her food, and nibble it from between my fingers as I held it close to my lap.

To my delight, this new way of feeding her worked at lunch ... and again at dinner.

I texted Lizzy with the news, and kept texting her after subsequent meals: "9th consecutive meal with no problem! Can hardly hear her swallow! Can hardly believe it!" I stopped counting and texting Lizzy after the 26th consecutive meal without any significant problems.

The 'new' way of giving Lily her food has now become the 'usual' way, as I'm still feeding her by making her stretch her neck.

Occasionally, she has startled me with a very loud, single cough, which happened because I'd lost my concentration when feeding her. I must stay focused. It takes only about seven minutes to feed her with my 'January method,' and the best part is that Lily feels a lot more comfortable at meals, too.

Lily has treats every day, but they must be very small. She is particularly fond of gingernut biscuits, and one biscuit provides four treats. I break the biscuit in half, then break each half into one larger and one smaller piece. Pilot gets the larger piece and Lily the smaller. Once, when the girls changed position behind my back, I gave Lily the larger piece, and she did a fair amount of coughing because of it. That hasn't happened again, I can tell you. Pilot wasn't too happy about having the smaller piece, either!

Store-bought dog treats must either be small or able to be easily broken into little pieces. High-value treats for obedience training consist of cooked chicken cut into pea-sized bits, or similarly-sized liver treats I make from a recipe that Jackie Foster, Pilot's breeder, gave me.

It's hard enough to make sure that Lily is able to eat the food she *should* have, but every walk carries the additional worry that she will find something she *shouldn't* have, try to eat it, and choke.

I could save a fortune on dog food if I fed her only twigs, and it's frustrating when she picks up twigs and crunches them. She hears me tell her "Lily, leave it," quickly followed by "Good girl" so often she probably thinks 'Leave-it-good-girl' is her second name.

Usually when Lily hears "Lily, leave it!" she spits out whatever she is chewing. If she keeps trying to spit it out and can't, I reach into her mouth and usually find a piece of a twig at the side of or under her tongue, or spanning the roof of her mouth. She doesn't always spit out what she's chewing, and sometimes 'Lily, leave it' is rapidly followed by her little pink tongue licking the sides of her mouth. Whatever it was, she had managed to swallow it. She doesn't hear 'Good girl' then.

Though twigs are bad news, she has picked up worse. When she was about ten months old, she found and tried to eat fox faeces on a walk one evening when we were almost home. She was choking on it and could hardly breathe as I pulled several handfuls of light brown, stinky, faeces-filled foam from her mouth. It was grossly unpleasant ... and it could have killed her. I'm very grateful that this phase of puppyhood is well behind her.

Like most dogs, Lily is attracted by the irresistible scent of food, thoughtlessly discarded in the street. Litter does more than stain our world with rubbish: it endangers the health and lives of animals. Pilot once plunged into a bush after scenting what turned out to be a discarded ham sandwich there. As quick as lightning, she began gobbling it down, but the problem was that the sandwich was still half-wrapped in clingfilm. Luckily, I managed to get most of the sandwich and all of the clingfilm out of her mouth. Had it been Lily who found the sandwich, it's frightening to imagine what might have happened.

Discarded, half-eaten burgers, rotting sandwiches, mouldy French fries, and almost-empty plastic yogurt containers are bad enough, but worse was the empty sardine tin with its lid still partially attached at a dangerous angle, ready to cut the tongue of any animal unfortunate enough to be attracted to it.

I've picked up all of the pieces of a smashed porcelain 'Birthday Boy' mug, and glass from shattered wine, beer, and alcohol bottles, on which dogs and other animals can cut themselves or ingest glass. For Lily, with her uncontrollable swallowing reflex, ingesting a small piece of glass would have dreadful consequences.

I am on constant alert for hazards when I walk Pilot and Lily, but at least they're with me so that I can watch them. If Lily were to go missing, I would have an even more serious cause for concern.

People who find a lost dog try to care for the dog kindly, whilst attempting to locate their owner. In an attempt to calm and reassure the animal, they may well offer food and water, which could prove fatal for Lily. I'd want her back as soon as possible so that I can feed and hydrate her properly, and have taken precautions in case of this eventuality. Lily has –

- a microchip (a legal requirement anyway)
- an ear tattoo, so she is on the National Dog Tattoo Register
- identification on her collar with her name, my postal code, and my mobile number

• a canine form of Medic Alert tag reading 'No food, no water. May choke to death'

Illness

Fortunately, Lily has never had aspiration pneumonia from food or sea water. Her only illness has been the occasional gastrointestinal upset, euphemistically referred to as 'runny tummy' by many dog owners. When Lily has a 'runny tummy,' she needs even more intensive, careful management.

For dogs without her condition, recovering from this ailment may involve nothing more than a dose or two of probiotic paste, and 24 hours without food (though with access to water). As mentioned previously, a dog needs a daily minimum of 30ml (1oz) of water for every 1kg (2.2lb) of weight. Lily weighs-in at 23.5kg (52lb), so requires at least 705ml (25oz) of water when well. But water is lost with diarrhoea, and so her water intake increases to about 1000ml (35oz) a day.

Lily gets most of her water from her soaked food. When she is not allowed to have food, I could let her lick lollies, but she'd need to lick 33 of them to get enough water to prevent dehydration. That's too many lollies to lick: she'd have no tongue left. I *could* break them up into chunks about the size of a five pence piece, as I do for her on a hot day when she's well, then hand-feed her a chunk at a time for her to chew and swallow. But her lollies are made from the water I drain from her kibble, so have tiny bits of food in them.

How could I hydrate her sufficiently if not with her usual lollies?

On the few occasions that Lily has been unwell, she has been seen by vets who had not previously looked after her, who were surprised when I told them how big a problem it is for Lily when she is restricted to no food. Not surprisingly, none of them had heard of her condition, much less treated a dog with it, and I needed to explain what it is.

Now, though, Lily has a dedicated vet, Dr Lucy Atkinson, who has taken it upon herself to learn everything she can about Lily's disability. She is well aware of the implications for Lily of 'No food for 24 hours. Only water,' and has prescribed rehydration support which, mixed with water, can be frozen into lollies for Lily. She has given Lily antibiotics intravenously rather than by mouth, and a probiotic paste to lick, which I smear in small quantities on the base of her lolly bowl.

When Lily was allowed soaked food again, Lucy prescribed tablets that I could crush and mix into Lily's food. I spooned the food/tablet mix – the consistency of very soft pâté – into her lolly bowl 5ml (1tsp) at a time for her to lick up, and she managed just fine. Thanks to Lucy, Lily was quickly back to her bright, tail-wagging self.

Lily will need intensive management all her life. If I had to make the decision again whether or not to help Lily survive, would I still choose to do so?

Yes. Absolutely.

I didn't decide to help her survive simply to prove I could. I helped her because I was so certain she would have a great life if she had the chance.

As I wrote about Liana earlier, life is beyond precious; once it is gone, it is gone forever.

We all love Lily deeply.

And Bentley would miss her big time if she wasn't around.

Big bro Bentley

Bentley is, quite literally, Lily's big brother, because he's her litter-mate, and he's also big, not just because he is older by 55 minutes, but also because he was big when he was born, the heaviest of the ten puppies that Pilot had, weighing 564g (20oz) compared to Lily's 406g (14.3 oz). And he's still big. At two years of age, he was a healthy 37.8kg (83.3lb) compared to Lily's 23.5kg (51.8lb).

He and Lily are exceptionally close, and they may actually always have been, even before they were born.

As you will have read in the first chapter, I kept extensive records about the puppies when they were born, but the one thing I didn't record, because I couldn't, was which 'side' each puppy came from. Until Pilot was scanned to show she was in whelp, I didn't even know that puppies could be on one of two 'sides.'

During Pilot's scan, I learned that the body of a bitch's uterus is very short, but has extremely long, narrow horns which give it an overall 'Y' shape. 20 to 21 days after the mating, the puppy embryos attach themselves to the walls of the two uterine horns, and grow like peas in a pod.

Around 40 days later, the puppies are born.

Recent evidence suggests that puppies are born in sequence from alternating sides of the uterine horns. The first puppy born was Liana, and the second – Bentley – arrived 65 minutes later, presumably from the opposite horn. The third puppy was Dodge, born 35 minutes after Bentley, who probably followed Liana down the same horn. The fourth puppy, Lily, was born 20 minutes after Dodge, in all likelihood from the same horn as Bentley.

One day, as Lizzy was reminiscing over our puppy birth record sheet, she made an astute observation: "Bentley was the second puppy born, and Lily was the fourth. Bentley came out hind legs first, and Lily came out front legs first. It's quite sweet, really, as they must have been nose-to-nose in the same horn waiting to be born. No wonder they love each other so much – they've always been together!"

It's fun to imagine that, as Pilot's contractions pushed Bentley toward the birth canal, he said to Lily, "Bye – see you in 55 minutes!"

As they grew up, Bentley was always around Lily, as a companion and sometimes a protector when the other six puppies may have played too roughly with her.

Now that they're both adults, Bentley and Lily still get along famously, and, after a hard afternoon of play when they fall asleep on the rug, some part of one is always touching some part of the other.

It's not always sweetness and light with them, though. Bentley does seem to be the catalyst for leading his mother and his sister into naughtiness, as happened the day he led them into a mud puddle. The three of them came out cream and black, filthy from the tummy down. I'm sure Pilot enjoyed the fun, but, in the photo

Lizzy took, she looks as though she's wondering why two of her puppies are still around when they were all supposed to be adopted.

The naughtiest thing Bentley has ever done was to eat my daughter's belt. We might have understood why he found it irresistible, had it been leather, but it was plastic. The buckle – which Lizzy found in the living room – was the only part of the belt that remained, and, after a few minutes of futile searching, she realized where the rest of the belt was: inside Bentley. She rushed him to the vet for an emetic, and he vomited it up in several pieces.

At least, Bentley hasn't encouraged Lily to eat a belt. Fortunate, really.

Best not to have a belt on her menu ... or on his.

Visit Hubble and Hattie on the web: www.hubbleandhattie.com
hubbleandhattie.blogspot.co.uk
• Details of all books • Special offers • Newsletter • New book news

One in 5983
(out of 8.5 million)

In May 2017, out of the 8.5 million dogs in the UK, just 5983 (0.07 per cent) were qualified Pets As Therapy (PAT) dogs.[1]

PAT dogs visit people in 9000 establishments, such as hospitals, hospices, residential and day care homes, and schools, providing vital comfort, companionship, and therapy.[2]

Assessment

To qualify as a PAT dog, an animal must have been with their present owner for at least six months, and be a minimum of nine months old before being presented for assessment. The assessment is understandably rigorous. I already knew what it involved because I took Pilot through it when she was ten months old. Though very young, Pilot passed every category with flying colours, setting the bar high for Lily. By the age of five, Pilot had made 200-plus visits to schools and residential and day care homes.

The assessment by a specially trained PAT assessor or veterinarian begins at a pre-arranged time and place unfamiliar to the dog. The assessor then undertakes the evaluation of the dog, and how suitable he or she would be as a PAT volunteer.

The points of the assessment are –

- **Assessor's initial impression of the dog's behaviour**

- **Assessor confirms whether the dog is presented on a collar and lead**
No slip leads, extendible leads, head-collars, harnesses or check-chains

- **Relaxed lead walking**
Does the dog pull on the lead at all, just a little, or a lot and strongly?

- **Control whilst talking to the owner when the dog is on a loose lead**
While the assessor talks to the owner, does the dog remain calm beside the owner, or require a few or many commands to do so?

- **Touching the dog**
Can the owner groom the dog's back, chest, stomach and tail whilst he or she

[1]From the well-respected Pet Population report commissioned by the Pet Food Manufacturers' Association.

[2]When Pilot began working in the Read2Dogs programme at a junior school, several little girls were surprised to learn that she is a girl dog. They told me: "Pilot's a boy's name." "No, it isn't," I assured them. "A girl can be a pilot. My daughter is a pilot. A girl can be anything she wants to be." It's nice to think that Pilot, just by virtue of her name, will help those girls consider high altitude flight plans for their future.

remains calm, or do they become excited, mouth the owner, jump, roll around or back away and seem reluctant to be touched?

- *Owner restricts the dog by holding their collar, and pulling them close*
Does the dog accept this or does he/she struggle, try to escape, or even show aggression?

- *Assessor and dog interaction*
When the assessor strokes and fusses the dog, is the dog calm and enjoying the attention, indifferent, or showing he/she isn't happy with it?

- *Assessor touches/examines the dog's paws, ears, and tail*
Does the dog happily accept this examination or does he/she show they do not want to be touched, especially in these areas?

- *Assessor offers a food treat between closed fingers*
Does the dog take it gently, greedily, not at all, or snatch it so that the assessor feels his/her teeth?

- *Assessor and volunteer remain seated for five to ten minutes*
Does the dog settle and wait quietly, sitting or lying down, or does he/she bark, whine, pull on the lead or nudge, demanding the owner's attention?

- *Without warning, the assessor makes a sudden noise by dropping something to simulate a situation where the dog may be working, and a walking stick, book or tray is dropped*
Does the dog look towards the object and then ignore it, investigate it, or bark and show fear?

- *The dog's general condition*
Does he or she move well and appear healthy, or require grooming, have bad breath, a dirty coat, or nails that need trimming?

- *Jumping up and pawing during assessment*
Does the dog jump up or paw at the owner or the assessor? How often, if so?

- *Other behavioural issues during the assessment*
Does the dog bark repeatedly, lick excessively, or exhibit any other behaviour that would make working as a PAT dog unacceptable?

The assessor's evaluation of Lily's behaviour on initially meeting her was 'Quiet and well-behaved. Naturally pleasant, and happy to meet people.'

In a way, Lily did even better than Pilot because, unlike Pilot, she had to cope with a distraction. During the ten minutes that the assessor and I sat and talked on a bench in a public park, where the assessment took place, three dogs came sniffing around and interfering with Lily, who remained relaxed sitting beside me, unfazed by their attention. The assessor was impressed, writing, 'Interested in her surroundings but remained focused on owner. Well behaved around other dogs.'

Lily – one in a million!

I knew I wouldn't be told the result of the assessment right away, as this had to firstly go to PAT head office, where the decision would be made. If Lily was accepted as a PAT dog, I would also have to provide the organisation with proof that she was up-to-date with her vaccinations. She would also need to be clean, as well as flea-, worm-, and parasite-free. One of her assignments could be to visit patients in hospital, or palliative care homes, after all.

Her result arrived by post a few weeks later. Lily had passed her assessment in every category with flying colours, just as her mother had, and she had been just nine months old when assessed. I was extremely pleased. My little puppy who had needed so much help to live would now be able to help others!

The certificate welcoming Lily as a registered PAT dog stated –

"Pets As Therapy visiting dogs and cats, and their owners, take on a commitment to provide a service to the community by sharing their healthy and temperament-assessed pets.

"The Pets As Therapy visiting scheme is accepted as therapeutic as well as bringing happiness and pleasure to people of all ages in establishments of various kinds, and is the largest scheme of its kind in Europe.

"Pets As Therapy dogs are recognized by the Royal College of Nursing, and welcomed by medical authorities in many areas."

It was a very proud and happy day when Lily first wore her PAT jacket, lead and collar, her photographic PAT identification tag hanging from it. Adults and children alike smile when I say in jest that Lily has her photo on her PAT ID tag so they'll know I've brought the right dog with me, and I have my photo on my PAT ID badge so they'll know that she's brought the right human with her!

Through her work as a PAT dog, Lily engages with people of all ages, from the very young to the very elderly.

Visits to schools

Lily works in the Read2Dogs programme when she makes her regular visits to two schools.

PAT limits dogs to working for one hour at a time, so Lily listens to four children read to her for 15 minutes each. The children might have difficulty reading, or they may be confident readers who have special educational, social or emotional needs, such as some form of autism, mutism, or speech, language and communication difficulties. They might be suffering from serious problems within their family.

During an initial Read2Dogs session, the children meet Lily and pet her if they wish. As part of a doggie show-and-tell, I explain that Lily was born unable to eat or drink normally. I take along some dry kibble and some soaked kibble, and maybe a frozen lolly in an insulated bag, and explain Lily's feeding and hydrating routines.

In each Read2Dogs session, the children sit on the floor as they read, Lily beside them, perhaps resting her head in their lap, or they might sit in a chair with her at their feet, listening. After 15 minutes, most of the children fuss over her and give her a treat – one at a time in case she chokes.

Lily also works with children who are seriously afraid of dogs. A phobia about dogs, a clinical condition called cynophobia, can make just walking along a street a terrifying prospect. When these children see a dog coming their way, they may cross the road to avoid the animal. Relaxing in a park may be impossible because dogs could be there. With 8.5 million dogs in the UK, it must seem to them as though there are dogs everywhere. Overcoming a fear of dogs is positively life-changing.

When Lily helps those who are very afraid of dogs, the initial session may involve nothing more than having them look at her from a distance, with a glass door between them.

As the sessions progress, the children choose by how much the distance between them and Lily should decrease. They may soon want to stand at the open doorway. They may take a few steps toward her. And, because she's a PAT dog, they can be certain that she won't jump up or rush at them.

Surprisingly soon, they usually decide to approach Lily, and even touch her somewhere on her body, though it's usually the end furthest from her teeth. (That's why one of the PAT assessments involves touching the dog anywhere on the body and watching their reaction.)

Thanks to Lily's calm, unthreatening nature, it took only three sessions for one dog-phobic child to ask to have his photo taken petting Lily.

One child on the autistic disorder spectrum benefited so much from Lily's regular, calming visits that his mother asked him what he thought had helped him to feel more settled. He told her there were five things –

- first, his personal classroom assistant
- second, me
- third, Lily
- fourth, his mother (who smiled at being fourth and took it very well)
- fifth, God.

One very anxious child benefited from Lily, not by reading to her, but by walking her around the school playing field, and throwing a ball for her, before praising her and giving her small treats.

The children chosen to take part in the Read2Dogs programme are usually in Year 4 (ages 8 to 9). But when a Year 4 school trip occurred on a day that Lily was scheduled to visit, so were not at the school, the children she had helped from the previous Year 4 came to see her instead, and read to her again to remind her of how much she had helped them to improve.

The school librarian wrote: "What a privilege it has been to be part of the Read2Dogs scheme. Lily's gentle and calm nature has encouraged our reluctant readers considerably. With no pressures or expectations, our readers have flourished, not only gaining in reading and literacy skills but also because of the noticeable and very positive effect on their confidence and self-esteem. Our pupils really look forward to Lily's visits, and thoroughly enjoy them."

The teacher at another school wrote:, "Lily's visits have brought numerous benefits to our children, that would be very difficult to achieve in any other way. Lily helps reluctant readers to enjoy reading again. Children who struggle socially feel completely accepted. And children who are afraid of dogs relax in her company. Lily's work is simply priceless!"

When Lily was spayed, her operation was scheduled for a week before the end of the school year in July so that she would not miss any Read2Dogs days, and would have the summer to recover. Four children at one school made a card for her: "To Lily, the lovely dog, We are sorry to hear that you do not feel like a 'Topdog'

Lily – one in a million!

right now. We hope you will get better soon because we all love and respect what you and Mrs Hamilton has *(sic)* done for us. We hope to see you and Mrs Hamilton back in September. Lots of good wishes from ..." and the children had each signed the card. The card came to my home, and was addressed to Lily, not me, so it was only good manners that Lily 'woofed' her reply, posted back to the children.

Visiting the schools is rewarding for both Lily and me. Lily loves being with the children, and I enjoy being back in a school environment, because I very much enjoyed my years teaching English, and it's extremely rewarding when I see the positive effect Lily has.

Visits to residential and day care homes

Lily helps residents with dementia in the two care homes that she visits. When they see and stroke her, she opens the way to memories of dogs who once enriched their lives.

Lily's visits mean a great deal to one particular blind resident, who, over the years, had had several Golden Retriever/Labrador guide dogs. Now, when he strokes Lily and holds out a treat for her, he smiles, reminiscing about his beloved guide dogs.

The Activities Co-ordinator at one residential care home wrote: "Lily has had a great impact on the residents and staff. She puts smiles on residents' faces. With her very kind nature, she is perfect for our residents, reaching to the bed bound who look forward to her visits. From her first visit, Lily has made our residents' day better just by being here. We look forward to Lily coming, and to her continuing her excellent work."

The manager of another care home wrote: "Lily is a delightful PAT dog with a kind and gentle nature. Her visits bring joy to the residents, often rekindling pleasant memories of pets they once had. The whole experience is always positive and therapeutic."

The residents in the homes really love Lily. On one extremely hot day, one resident, Elizabeth, was so concerned for Lily's comfort that she waited to see Lily not in her bedroom, as she usually did, but in a lounge, because it was cooler for Lily there. At this same home, a resident whose name was also Lily, enjoyed seeing Lily and petting and stroking her so much that she told her family about it with smiles and great enthusiasm. Lily's visit was one of this woman's last lovely memories, for she died two weeks later. Her family told me how much it still means to them that Lily brought such joy to their mother in her final days.

PAT poster girl

When Lily was 11 months old, she and Pilot became poster girls to raise money for PAT at my local Waitrose, in the 'Waitrose Community Matters' enterprise. Every month, the larger Waitrose supermarkets divide a donation of £1000 among three charities. Each time customers shop, they 'vote' for the charity of their choice by dropping a green token in that charity's box.

Of the £1000 donation Waitrose gave to charities that month, Pets As Therapy received £506 of it.

By the good that she does as a PAT dog, Lily is ensuring that her presence in this world is especially worthwhile.

Not the end of her story

Lily has been and continues to be an amazing dog. To look at her, you wouldn't think there was a thing wrong, and might not even appreciate that she is somewhat small for a Golden Retriever.

When Pilot was in whelp, I looked forward to having puppies, but I also looked forward to the day that it was just Pilot and I again. We were a great team. She was eight weeks old when she came into my life, four months after Jim passed away from cancer. She has helped me survive my bereavement thus far. But while I'd never intended to have two dogs, I have two now because of Lily. And I wouldn't want to be without her.

Learning to help Lily live, and learning to live with Lily has been challenging, demanding, rewarding, fulfilling and, ultimately, successful.

Lily will always need to be fed carefully. I will have to soak her food to ensure she ingests water with every mouthful; be vigilant and attentive, listening to her every swallow; watch her body language, and be alert to gurgles and grumbles in her throat so that I can pause feeding her before she gets into more trouble ... unless she wanders away for a break and a cough first. But at least my most recent method of feeding her has made life easier for us both.

As long as she can be prevented from aspirating food and developing pneumonia, she should live as long as any normal Golden Retriever.

Lily greets every day full of the joys of life, happy to be alive. She has a lot more running around and mouth-wrestling to enjoy with Bentley and Pilot, and loads of doggie games to play. More tail-wagging, balls to retrieve, walks to enjoy, squirrels and deer she can only dream of chasing because chasing is never allowed, and the inexplicable pleasure she derives by supervising the way I load the dishwasher; adding her own pre-wash touch with precision licks.

When she's not helping with the dishwasher, working as a PAT dog or having fun with Pilot and Bentley, she enjoys playing sports.

She's astonishingly good with tennis balls. If she had opposing thumbs, she could be Wimbledon's first canine champion. Her way of jumping into the air to catch tennis balls is definitely impressive. Even more impressive is the way that she plays with a tennis ball. Lizzy, two of my friends and I have watched her several times lie on her back, roll the ball around with her front paws, throw it from one paw to the other, slide it along her front legs, mouth it, then roll it back along her front legs, toss it again, and catch it with her paws in the air.

One of my friends, Leah, who watched Lily with the tennis ball recalls "I do remember her doing this. We were entranced and couldn't believe Lily could be so dexterous with her paws."

While Lily enjoys tennis balls, she absolutely loves footballs, and is so good with one that her human opponents are rarely able to take it from her. I wouldn't be unduly surprised if a major football team didn't want to sign her up! She's ready, but

Lily – one in a million!

only as long as she can play by her own rules, one of which allows her to carry the ball in her mouth.

One sport Lily has never wanted to try is swimming, and never goes into the sea any deeper than up to her tummy, almost as if she understands that liquids are potentially dangerous for her, because a mouthful of seawater while she's swimming could end up in her lungs, and the result might be aspiration pneumonia.

We in her human family, other people who know her, and the many adults and children she has helped and continues to help are delighted she is still with us. Bentley most definitely is delighted, for she's his best friend and playmate.

I didn't need to turn Lily into a little angel. She did that herself by helping so many, making her presence on this planet count, and surpassing all my hopes for her.

Lily really is one in a million and a miracle of survival. She could so easily have lost the gift of life after she was born.

With so much still to enjoy, she's not stopping now.

Lily's future is wonderfully bright.

Visit Hubble and Hattie on the web: www.hubbleandhattie.com
hubbleandhattie.blogspot.co.uk
• Details of all books • Special offers • Newsletter • New book news

Lily's day

The most important part of Lily's 'day' must actually take place the night before, otherwise, the next day will definitely not begin well: her food won't be ready.

10:00pm

Pilot and Lily are snoozing in the living room as, in the kitchen, I begin measuring 140g (5oz) of weight-control kibble into each of Pilot's two bowls. The sound of the kibble biscuits bouncing into the bowls wakes the girls. Within seconds, they've come to watch. And wait. Into each of Lily's three containers, I measure 86g (3oz) of adult medium-size dog kibble.

When I've finished, their waiting pays off, as I feed them each a few pieces of dry kibble. Then Lily trots into the garden for her last comfort break of the day. Pilot doesn't need one – she has long-range tanks.

After letting Lily back in, I pour 500ml (20oz) water over Lily's dry kibble almost to the top of the containers, click on their lids, and put them in the fridge for the kibble to soak overnight.

Back in the living room, Lily curls up in Jim's chair. Pilot stretches out on the couch beside me while I watch TV and charge their flashing collars for morning.

6:15am

I get the girls ready for the day's first walk. It's below 5°C (41°F) so I put their coats on them. They look unhappy about this, but I tell them they'll thank me when we come home and their backs and tummies are warm.

On go their flashing collars: Lily's twinkles pink; Pilot's twinkles green. Better for them to be seen than not, even though only four cars pass us.

6:50am

We're home after quite a cold walk. Taking off the girls' coats, I pat their lovely, warm backs, and tell them how lucky they are. I'm not sure they're convinced, though Lily wags her tail a little.

7:00am: breakfast

Lily's kibble has absorbed 215ml (7.5oz) of water, and each piece of kibble has swelled from the size of my little fingernail to bigger than my thumbnail. I warm her breakfast in the microwave for 70 seconds to take the chill off it. The microwave dings to let me know that her meal is ready.

I drain the liquid from Lily's food, and use half of it to moisten Pilot's, saving the rest to make some of the dozen or more lollies that Lily will need during the day.

After giving Pilot her food, I sit down, protecting my lap with two tea towels and a kitchen towel. Hand-feeding Lily is messy because her soaked kibble is slippery.

Lily – one in a million!

Spacing her mouthfuls five seconds or more apart, I automatically count them as I feed her. I'm still relieved when we pass mouthful 17 without any coughing, choking or gurgling. My new method of feeding her makes her mealtimes almost always choke-free – a huge relief for us both!

7:15am

After finishing breakfast, Lily stands by Pilot's empty food bowl and looks at me. She wants lollies. Taking three from the freezer, I splash them with water, arrange them in her lolly bowl, and place it in Pilot's empty food bowl. Pilot's bowls are in a raised feeder 28cm (11in) high, so Lily doesn't need to bend down to lick her lollies. I listen to hear if she begins to gulp air as she licks, and, if she does, I'll take the lollies from her. But she's fine.

Pilot, as usual, gets treats when Lily has lollies.

9:15am

Today is one of the two days each week that Lily works at schools in Pets As Therapy's Read2Dogs programme. Pilot stays at home, happy with a roly-poly, treat-filled toy. When Pilot worked at this school yesterday, Lily went to Lizzy and Shaun's to spend the hour or so with Bentley, because, unlike Pilot, Lily can't have a toy with treats in it to occupy her when alone.

11:30am: walk-time

In the nearby woodland, there are sticks for Pilot to carry and, unfortunately, twigs for Lily to chew. When she picks up twigs, I tell her "Lily, leave it!" and she obediently spits out the twig.

Off-lead at the grass recreation ground, Lily jumps up to catch a ball as it bounces high while Pilot rolls around. Then they join three doggie friends and chase around for ten minutes.

Lily has two lollies when we get home.

12:15pm: lunch

Lily needs to have her food divided into three daily meals to ensure she's regularly hydrated. Pilot thinks this is great, because she gets a 20g (0.7oz) snack from her dinner now so that she doesn't feel left out when Lily has lunch. I measure Pilot's while Lily's lunch warms for 70 seconds in the microwave.

Lily has another successful meal. No choking, no coughing, and finishes with three lollies.

1:30pm

Bentley arrives because, today, both Lizzy and Shaun are working an afternoon shift, and can't be with him. The three dogs enjoy hours of play, chewing naturally-shed stag antlers, and rearranging my throws and cushions.

2:30pm

Lily stands by Pilot's food bowl and looks at me, telling me she's thirsty. As I get lollies for her, Pilot and Bentley hear me open the freezer, and appear, almost by magic, because they know that this sound means lollies for Lily, and treats for them.

I freeze more lollies now using the water drained from Lily's lunch, and then go out for a couple of hours. The dogs watch me reproachfully. How can I leave them?

4:45pm
I arrive home to see all three dogs at the window, watching for me. To show how happy they are that I haven't deserted them forever, they each bring me a toy. After five minutes, they go out into the garden for a run around and a comfort break.

5:15pm: dinner
Lily has a dozen lollies already in the freezer so I use all the water from her dinner to moisten the kibble for Pilot and Bentley.

I put down Pilot's and Bentley's food and they tuck in, then I sit down and feed Lily. After Pilot and Bentley finish, they check each other's bowl for kibble. Finding none, they stay, watching Lily eat but never bothering her. After her meal, Lily has three more lollies, so Pilot and Bentley have treats.

After my dinner, Lily hinders – or 'helps' – with her own interpretation of pre-wash as I load the dishwasher.

7:30pm
I take the girls for their last walk of the day. It's dark and cold so they wear their coats and flashing collars. We're back in 15 minutes, and then I take Bentley for a short walk.

9:00pm
Lily and Pilot try and ignore Bentley's invitation to play, but his persistence finally pays off and the three engage in tug-a-toy, mouth-wrestling, and harmlessly chewing on each other's ears and legs. Then Lily chases Bentley around the dining room table.

9:30pm
Lily and Bentley are asleep on the living room carpet, her head on his shoulder. I unload the dishwasher, and begin measuring the kibble into the bowls for tomorrow. The dogs wake at the sound and come to watch and wait for treats.

Lily and Bentley charge into the garden for the day's last comfort break. Pilot stays inside.

After I let Lily and Bentley back in, all three watch me add water to the kibble in Lily's containers, click on the lids, and put them in the fridge. Lily has one last lolly of the day. Pilot and Bentley have their last treat of the day.

9:50pm
Lizzy arrives to collect Bentley, and this excites all the dogs because she fusses over them.

Minutes after Lizzy leaves with Bentley, Pilot and Lily are asleep in their usual places: all's quiet for the rest of the night.

Another typical day with Lily draws to a close ...

Index

More from Hubble & Hattie ...

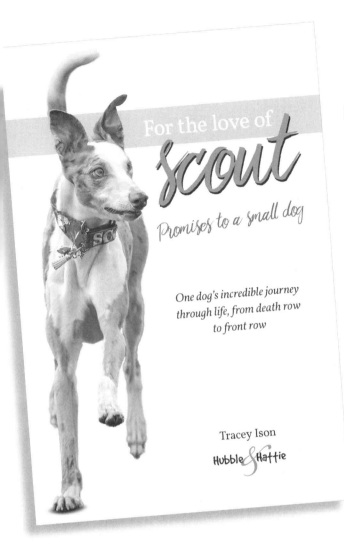

"... teaches us about love, and the resilience of canine and human hearts ... This book shows how one can repair broken hearts with gold so that they are stronger, and shine brighter than before"
– Geelong Obedience Dog Club

"Scout's stories are unique and delightful, but also brought me to tears on occasion ... A lovely story, a beautiful, true-life book, with two generous sections of photos of Scout, his family and friends"
– Dog Training Weekly

Bringing light and hope to the dark world of an unwanted stray ...

Rescued from almost certain death as a puppy, Scout, the short-sighted Lurcher, began to understand just how it felt to be loved.

Join Scout, his family and friends, as they face the daily challenges that living with a partially-sighted sighthound involve.

Paperback • 112 pages • 47 colour pictures
• ISBN: 9781845849368 • £8.99*

For more info on Hubble and Hattie books please visit www.hubbleandhattie.com;
email info@hubbleandhattie.com; tel 44 (0) 1305 260068
*prices subject to change/p&p extra

"An inspirational book full of true tales demonstrating that if you love dogs, you will know they have the power to change lives. Extraordinary, powerful and compassionate"
– Dog Training Weekly

" ... an exploration of nurturing positive interactions between two unique species"
– Geelong Dog Obedience Club

HOUNDS WHO HEAL
It's a kind of magic!

CHRIS KENT
Foreword by Dr Daniel Allen

Hubble & Hattie

The desire for human connection is a fundamental need. For some, however, the closest they come to this connection begins with a dog.

This is the story of six abandoned dogs, who ended up living together and inspiring the development of the unique K9 Project.

It's also the story of the people they met, the ones they helped, and the ones they couldn't ...

Paperback • 128 pages • 56 colour & 34 mono pictures
• ISBN: 9781845849733 • £10.99*

72